MW00452172

The Current

New Wheels for the Post-Petrol Age

gestalten

LIGHTNING IN A BOTTLE

How the Best Idea of the 1800s Took a Century to Realize

Electricity is modernity. More than any other technology, the generation and harnessing of electricity has transformed the daily life of human beings around the planet. Access to electricity, still a challenge in parts of the world, is the barrier between a hundred thousand years of human history, and every-thing afterwards: it broke our bond with the sun, the stars, and the horse, making us appear independent of natural cycles. It was the true Promethean moment. Optimists see a future that is increasingly entwined with electricity, an escape hatch from a century of dependence on the filthy, infernal petroleum.

We side with the optimists in this book, exploring the history of electric vehicles and their present-day blossoming. We enthusi-astically embrace the logic and promise of relatively clean energy production, the use of an existing distribution grid, and the replacement of a problematic technology with a superior one. The electric option was a conceivable path over 100 years ago, but the world was too impatient to wait for clean, reliable, and fast electric vehicles. We are paying for the geopolitical and environmen-tal consequences of petroleum every day, and we must wonder how the world might have been different had we chosen lightning over fire in 1900. Fire was not stolen from the gods—it was always present on earth. But electricity is a child of the sky and the air. Channeling lightning borders on magic, while containing small explosions through a system of lubricated chambers and joints, spewing noise and poisonous smoke, seems increas-ingly like a Faustian bargain.

The Origin of the Species

The development of electric motors began in 1821 with Michael Faraday, who laid down the principles that all subsequent e-motors would follow. The rechargeable lead-acid battery was invented in 1859 by Gustave Planté, and the first dry-cell batteries, like the ones we use in flashlights, were invented in 1886 by Carl Gassner. Batteries, always the most problematic component of electric vehicles (EVs), were initially built in a bewil-dering variety of types: dry and wet, closed- and open-cell. Still, their potential for powering vehicles was clear from the start.

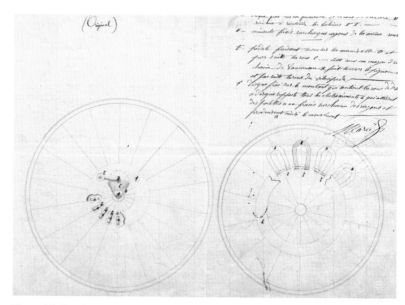

The world's first electric bicycle? Guillame Perraux's 1868 patent for an electric bicycle motor.

The first concept for an electric motor-cycle came from the same man who first patented the motorcycle itself, Louis-Guillaume Perreaux. His 1870 steam veloci-pede was likely the second motorcycle ever built (the first being Sylvester H. Roper's 1869 steamer), but Perreaux's 1869 patent drawing does not specify the motive power. In 1870 he patented his steam-powered cycle, along with a design for an electric motor capable of powering the cycle. It is an elegant drawing, but nobody knows if he actually built the electric motor, or whether he ever attached it to his velocipede.

Pioneer electrical inventor Gustave Trouvé's 1881 e-trike, the world's first EV.

It is generally accepted that Gustave Trouvé, the lost pioneer of electric technol-ogy, built the first human-bearing electric vehicle. He demonstrated his eccentric tricycle on April 19, 1881, along the Rue Valois in central Paris, attaching his own batteries and electric motor to a Starley tricycle. He could not add the electric car to his list of 300 patents, as a rival had already patented a Humber trike with a steam engine, so Trouvé hooked his motor and battery up to a propel-ler and installed the unit onto a boat—thus the outboard motor was born.

In 1885 John Kemp Starley built the first successful Rover Safety Bicycle. With this bike, he laid the foundation for both the bicycle and motorcycle industries as we known them. All rolling vehicles improved dramatically with Joseph Dunlop's inflated rubber tire of 1890: balloon tires cushioned road shocks and led to an enormous bicycle boom in the United States, the United Kingdom, and Europe. The sport of bicycle racing became hugely popular, with banked-track velodromes sprouting up in nearly every town. Colonel Albert Pope of Columbia Bicycles was the first to recognize the entertainment value of powered two-wheel-ers; he commissioned Sylvester H. Roper to attach a steam motor to a Columbia frame. The result was the fastest motorcycle in the world, reaching 42 mph (68 km/h) in 1896. Internal-combustion (IC) motor pacers came next, but they were notoriously unreliable until 1900. Electrically powered pacers were a natural solution. By 1898, Thomas Humber had added batteries to his pacing tandem, and he was not alone: that year 17 electric tandems competed in the World Championships. This was likely the first time racing spectators had ever seen a motorized vehicle.

Eugen Dutrieu's self-built 1898 electric tandem pacing cycle.

Meanwhile in Paris, Camille Jenatzy was busy proving the potential of electric automobiles; on May 1, 1899, he piloted the first vehicle of any kind to surpass 60 mph (100 km/h). The vehicle, named La Jamais Contente (French for "never satisfied") used aluminum torpedo-shaped bodywork atop a heavy chassis. Jenatzy sat half outside the sleek bodywork: it was not actually streamlined, but it was very modern-looking. With a pair of Postel-Vinay 25 kW motors driving the rear wheels, it produced about 68 hp, drawing 124 amps at 200 V, and the massive undercarriage of the svelte missile concealed the sheer weight of the batteries required to deliver such power.

Camille Jenatzky's 1899 torpedo speedster, "Le Jamais Contente", the fastest vehicle in the world before 1900.

The 1900s: One in Three on the Road

At the beginning of the twentieth century, electric cars outsold gas-burners and steamers in the United States. By 1900, New York City had a fleet of 60 electric taxis, and one third of vehicles on public roads in the United States were electric. They sold well for another 10 years; they were easy to drive, and since the majority of automobiles were driven in urban areas, range was not a concern. Electric cars were especially popular with women drivers because handcranking an IC car engine to life was an occasionally arm-breaking business. Even Thomas Edison, the American guru of all things electric, built a battery-powered front-wheel-drive electric car in 1895, and in 1912 made three more refined prototypes. For publicity—perhaps more for his batteries than for his car—he drove from Scotland to London at 25 mph (40 km/h), covering a remarkable 170 mi (274 km) stretch without recharging. Edison was convinced that eventually "all the oil would be pumped out of the ground." Electricity was the future.

While Henry Ford's Model T is generally credited with killing the electric and steam car markets, Ford revered Thomas Edison, and in 1914 they teamed up to design an electric Edison-Ford car. Ford claimed, "Within a year, I hope, we shall begin the manufacture of an electric automobile." But after a year of development, $1.4 million

(€1.2 million) invested, and a lot of press releases promising imminent production (still a PR habit of the EV industry in the twenty-first century), the project was abandoned because Edison's nickel-iron batteries were problematic. But this is not why electric cars disappeared from the roads; it was another electric motor—a traitor of sorts—that killed the electric car by the 1920s: the Delco electric starter.

Electric motorcycles were rare in the early years of the twentieth century, but their simplicity, silence, ease of use, and low maintenance made them catnip to home tinkerers. They appeared in a continuous stream of homemade inventions from the 1890s onwards, in press and in patents, but their limitations were clear: low power and short range. But changing economic conditions would highlight the appeal of electric transport, especially when petroleum was expensive, rationed, or even impossible to buy.

The Dark Ages

Electric vehicles found important niches in the heyday of internal combustion: utility, public transport, and as a backup technology. Electric vehicles were popular for donkeywork: hauling equipment, pulling boats through canals, and tugging planes. From the 1920s onward, private electric vehicles on two wheels or four were rare. But in dark times, we turned to electricity for help; the Depression years of 1930–1936 gave a boost to electric motorcycles when several manufacturers building e-bikes and e-scooters sprang up in Europe. French moto manufacturer Favor and the Dutch Simplex both sold sensible electric bicycles from 1932, with a large lead-acid battery pack at the bottom of the frame and a mid-mounted engine driving a secondary chain to the rear wheel.

In 1936 T. Hart installed batteries and a 12 V motor into an ABC chassis, a homemade cure for the Depression.

Despite their lackluster performance, EVs were still considered the motor of the future in the 1930s. In 1936 *The Motor Cycle* tested what was perhaps the first mass-produced electric motorcycle, built by the Belgian firm Socovel. By then over 1,000

Socovels had been built, prompting *The Motor Cycle* to ask the question: "Has the electric motorcycle a future?" While its speed and range were limited, it "might attract many whose needs are not met by motorcycles: no gears, no clutch, no starting difficulties—merely a twist of the grip on the right handlebar and the machine glides silently away. Could anything be more simple?" The Socovel was representative of e-bikes of the lead-acid battery era through the 1970s, but was still "a gentlemanly machine if ever there was one."

The war years of 1939–1945 spurred a wave of electric motorcycles developed to circumvent fuel scarcity in Europe and the United States. Most were homebuilt in desperate times. In the case of Merle Williams of Long Beach, California, it was fuel rationing that spurred him to convert his personal motorcycle to electric power. Demand for replicas inspired him to found the Marketeer Company, which eventually grew to build electric bicycles, small motorcycles, and, from 1951 on, electric golf carts. Golf carts became the popular interface with private electric vehicles: their ease of use and limited performance shaded everyone's view of EVs for decades to come. It took the sport of electric drag racing and, ultimately, the Tesla to erase their dowdy image.

The mass production of transistors in the 1960s shrank everything that handled electricity, from radios to computers, and eventually, electric-motor power controllers. It took decades for the EV industry to adopt tangential technologies for cars and motorcycles, but after a stagnant period spanning the 1950s–1970s, a combination of political events and government support gave a real push to the industry. By 1965, neodymium rare-earth magnets transformed electric motors, making them exponentially more powerful and reducing their size. We can also thank the space race for spurring new battery development: while it did not help the performance image of EVs, NASA's electric Lunar Rover, the first manned vehicle to drive on another planet, was geek-sexy to inventors.

The Oil Crisis and NASA

Electric vehicles in the 1960s and 1970s were everywhere: forklifts, golf carts, utility carts, postal delivery vehicles, and trikes for traffic wardens—none of which seemed a harbinger of an exciting electric future. But the 1973 Arab Oil Embargo, and the resulting oil crisis, put a spotlight on the geopolitical issues of petroleum. EVs were jolted back to life. Government funding for research was increased in the United States when Congress passed the Electric and Hybrid Vehicle Research, Development, →

and Demonstration Act of 1976. The environmental movement was at full strength, and the 1963 Clean Air Act was upgraded to include motor vehicle emissions in 1970. California led the world in pollution control, playing a vital role in promoting Low Emission Vehicles (LEVs) and Zero Emission Vehicles (ZEVs) through regulation, which

Then again, not everyone needed to do 100 mph (161 km/h) on two wheels. Just as in 1900, the needs of urban commuters were modest and easily fulfilled by existing electric technology in the mid-1970s. Several factories entered the post-oil-crisis electric market, such as Panasonic of Japan, Motobécane of France, and Hercules of Germany, each

The future too soon, or a bad idea? The Sinclair C5 was the largest mass-produced EV of the 1980s.

Mike Corbin's Quicksilver, shockingly fast at 165 mph (265 km/h), made possible by Yardney's NASA batteries.

scared the IC industry into action, for better and for worse.

Even before the oil crisis, a visionary named Mike Corbin saw the potential of the electric motorcycle and built outrageous speed machines to prove it. Corbin built his first land speed-racer in 1972, the Quicksilver, which used normal lead-acid batteries to power a pair of jet-engine starter motors salvaged from a 1950s Douglas A-4 Skyhawk fighter. His homebuilt semi-streamliner chassis was big and heavy and crude, but it was also the first e-moto to break the 100 mph (161 km/h) barrier on the Bonneville Salt Flats. Quicksilver was the fastest electric motorcycle to date, using technology available in 1910.

In 1974 Corbin teamed with Yardney Batteries, which had developed powerful silver-zinc batteries for NASA. This was literally outer-space technology: such batteries were powering both Sputnik and the Lunar Rover, plus nearly every satellite and off-world vehicle since then, because they offered eight times the energy density of lead-acid batteries. With Yardney supplying Quicksilver with NASA juice, the revamped streamliner clocked 165.397 mph (266 km/h) at Bonneville, an electric motorcycle world record that stood for 35 years! A new generation of tinkerers took notice: 165 mph (265 km/h) was no joke.

offering stylish electric bikes and scooters. They banked on the same appeal EVs have always had—silence, simplicity, and cleanliness. Then they combined it with the emerging environmental consciousness in order to sell the concept of clean energy transport.

Yamaha surprised itself in 1993 with the tremendous success of its pioneering pedelec (pedal-assisted electric) bicycle. The PAS (power assist system) e-bike gave riders a boost from its small lead-acid battery that only lasted 12.4 mi (20 km) and took hours to recharge, but the world wanted it! Yamaha sold 30,000 units of the PAS in the first year, which was triple its forecast. Major updates to Yamaha's PAS bicycles continue to this day, and over four million have been sold.

Sir Clive Sinclair, a personal computer pioneer, was the first person to combine computer technology and electric vehicles. Leveraging his vast fortune to invent a new transport category (the personal electric vehicle), he revealed, with much fanfare, the Sinclair C5 in 1985. It was an electric-boost pedal tricycle with smooth plastic bodywork designed by Lotus. It looked like the future, but its 15 mph (24 km/h) top speed, and worries about real-world road safety, meant Sinclair's expected transport revolution was met with sales of 5,000 units, with 9,000

sitting unsold before the company went bust. That still made the Sinclair C5 the biggest-selling electric trike in the world.

A Quiet Revolution

If there is a dividing line between then and now, a line separating the century of struggling electric technology and the sleek, fast present, it is the year 1995. The whirring of a cordless electric drill was the clarion call announcing a revolution in electric vehicle power. Power-dense batteries using lithium-ion (Li-ion) and nickel-cadmium (NiCad) were mass-produced for portable power tools and cameras, which lowered their cost significantly. Also, the increasing availability of rare-earth magnets made much smaller, more powerful electric motors possible. The combination of powerful, compact, and inexpensive batteries with

General Motors EV1 was an unexpected success, with bigger demand than the existing 1100 cars.

smaller, yet more powerful, motors was a watershed moment for electric vehicles of all kinds, and the mid-1990s is the dividing line: everything that came before was relatively crude, and everything that came after defines the contemporary era.

One of the best-known EV sagas of the 1990s was the General Motors EV1 program, which highlighted the tension between an entrenched IC industry and the increasing viability of EVs—even within the same company! The EV1 story began in 1987 at the very first World Solar Challenge, an impressive event where solar-powered vehicles from all over the world raced across the continent of Australia, traveling all the way from Darwin in the north to Adelaide in the south. GM's Sunraycer, jointly developed with Hughes Aircraft and AeroVironment, won the inaugural race. Next, GM and AeroVironment developed the Impact, an EV sedan unveiled at the 1990 LA Auto Show. The project that followed, the GM EV1, revealed unexpected public enthusiasm for the electric car, and the auto industry's horror at its popularity. In 1994, GM's initial call for 50 volunteers to test the EV1 in LA and NYC was met with over 10,000 requests in each city. *Motor Trend* said in 1996, "The Impact is the world's only electric vehicle that drives like a real car." People loved their EV1s, and GM leased over 1,100, but by 2002 they had crushed all but 40. The IC auto industry had also crushed California's law requiring every manufacturer to supply EVs, modifying it to accept hybrids and other low-emission cars. A terrific documentary, *Who Killed the Electric Car?* (2006), was both a greatly exaggerated obituary and a study of the enormous global forces aligned against EVs.

In 1999, Honda released the world's first hybrid-electric car, the Insight, but the global release of the Toyota Prius in 2000 transformed the auto industry. With celebrities lining up to buy them, and limited initial supplies, Prius mania changed the public's view of EVs as desirable and coveted. It would take the release of the Tesla, though, to make them sexy.

The E-Scooter: the Most Popular EV in the World

Peugeot's Scoot'Elec was the first mass-produced new-tech electric scooter, using NiCad battery packs. Peugeot's investment was prescient; over the next few years, electric scooters would become the most-produced EVs in the world. In 1999, China classified electric scooters as bicycles (no license required) if they adopted a 20/40 rule: 12.4 mph (20 km/h) max speed, 88 lb (40 kg) max weight. Lax enforcement of these rules, combined with a total ban on urban IC scooters in the mid-2000s, saw manufacturers racing to fill the transportation void left by the new law with cheap e-scooters, which now number up to 200 million units. This made China the largest adopter of EVs in the world, highlighting the dance between government regulation and the EV industry, which is ready to blossom with a stroke of a legislative pen.

With the popularity of the Prius and the excitement generated by the Tesla Roadster in 2008, the time was ripe for high-performance electric motorcycles. The e-scooter company Vectrix displayed a prototype sports e-moto with a claimed top speed of 125 mph (201 km/h) in 2007, but the most

The world's first e-superbike from the hand of Yves Béhar and the team at Mission Motors: 150 mph (241 km/h) guaranteed in 2009.

compelling design came from the team at Mission Motors, who partnered with industrial design wunderkind, Yves Béhar, to produce the Mission One in early 2009. It was the world's first electric superbike, mixing superb performance with groundbreaking design. The Mission One recorded 150 mph (241 km/h) on the Bonneville Salt Flats and competed in the new Time Trial Xtreme Grand Prix (TTXGP), taking fourth place in the hotly contested Isle of Man. The Mission One also won at Laguna Seca in record time and set an electric street-bike record on the drag strip with a blazing-fast 10.602-second quarter-mile time at 122.602 mph (197.3 km/h). The ultra-high-performance e-moto had arrived.

What brought electric motorcycles to Tesla-level performance was a wide-ranging mix of new battery technology, stronger and more compact motors, reliable and sophisticated programmable controllers, and the social interactions of an internet community of experimenters sharing information. The advent of computer-aided design (CAD) programs, computer modeling, and inexpensive 3D printing made breakthroughs inevitable, and the rapid transformation of the electric motorcycle into a wicked-fast beast exemplified a twenty-first century model of technological innovation. And yet, the story was virtually ignored by the traditional motorcycle community and press, who simply didn't—and still don't—seem to get the appeal of electric bikes.

From 2010 onwards, the EV industry kicked into high gear. Every major European motorcycle company began developing e-motos, and even the stalwart Harley-Davidson toured a fleet of LiveWires to dealers in 2014, salvaging a big gamble on

battery advancements that had not yet occurred. That the world's most conservative motorcycle company thrust its hand into the EV game was an oddity to H-D traditionalists, but revealed EVs as vital for future planning. H-D made this clear with the news of their major investment in Alta Motors in 2018. While rivals BMW and Piaggio have produced EVs for years, the full-size motorcycle market is one of the last dominos to fall in universal EV acceptance, along with 4 × 4s (and we have one in this book!). High-performance e-motos like Energica's Ego (the basis of a new Moto-e GP series) and Alta's podium-ready Redshift motocrosser are customer-ready versions of competition EVs heralding electric dominance: Carlin Dunne's Lightning that won Pike's Peak in 2013, the KillaJoule land speed-racer hitting 270 mph (435 km/h) at Bonneville, and even the Volkswagen I.D. e-racer that obliterated four-wheeled competition at Pike's Peak in 2018.

The saga of electric motorcycles is rapidly evolving, and this book is full of inspiration. Despite traditional motorcyclists' resistance to electricity, the sheer creative energy visible around e-motos is an indication that the technology will move forward at a rapid pace, winning converts with a mixture of good design, great performance, and a healthy dose of fun. While electric vehicles have always made excellent beasts of burden, the joy factor that was hinted at 100 years ago—the zippy electric future—is finally here. ●

Paul d'Orleans is a journalist and expert on motorcycle history and culture, and publisher of TheVintagent.com.

Surfer's Delight

When you don't need a gas tank, how to fill the void? Developing an electric bike means resolving design challenges, and Martin Hulin and Pierre-Yves Gilton of Essence Motocycles answered that question in a spectacular way: the plywood seat, supported by a tubular-steel trellis frame, bends to create the iconic shape of the gas tank, highlighting its absence. The wooden seat is reminiscent of a surfboard—or a vintage plywood Eames chair, depending on your preference. And while a chair is static, the surfing analogy remains true: "The frame geometry was designed for a natural ride whatever the riding mode. The feeling is crisp and sharp," explains Martin Hulin in an interview. The aptly named E-Raw stretches the limits of change: it takes its inspiration from classic custom bikes and café racers, but it beams them into the future of personal transportation. With specifications such as a 124 mph (200 km/h) top speed, a 124 mi (200 km) range, and a 30-minute charging time, this seems a carefree—but essential—electric experience. Interested? Be quick, because only 10 bikes in total will be built.

Acceleration
0–62 mph (0–100 km/h):
< 3.5 sec

Range
95–124 mi (153–200 km)

Power
109 hp (80 kW)

Torque
210 N m (154.9 ft-lb)

Top Speed
124 mph (200 km/h)

Battery
12–15 kWh battery pack

Weight
368 lb (167 kg)

Availability
Limited edition of 10

Three Wheels Good

The future of cute! Estonian carmaker Nobe must have a psychologist on its design team because, as every advertising expert knows, cuteness can go a long way. And it certainly works with the electric trike Nobe 100, bringing back memories of the 1950s, when people fell in love with microcars such as the Goggomobil, the Isetta, or the original Fiat Cinquecento. As a three-wheeler and three-seater in a tadpole configuration, this car enjoys improved aerodynamics and efficiency. And nowadays, this configuration also means improved handling; each wheel comes with a built-in motor creating a kind of 3 × 3 AWD. The dual-battery system provides a combined range of 137 mi (220 km), and one of the batteries is portable to help relieve range anxiety. The lightweight, composite body comes with a targa top (semi-convertible with a removable hardtop). The minimalist, classic interior also adds to the nostalgic ambiance of a summery mountain meadow dotted with flowers. If it was British, you could call it "quite a character"—and that would be meant as a compliment.

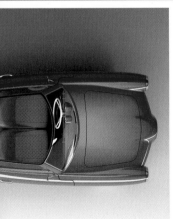

Acceleration
0–62 mph (0–100 km/h):
less than 6 sec

Range
137 mi (220 km)

Torque
490 N m (361.4 ft-lb)

Top Speed
68 mph (110 km/h)

Battery
21 kw/h Li-ion

Weight
860 lb (390 kg)

Availability
Pre-production

The Spy Who Loved Me Back

Coming to a spy movie near you! BMW Motorrad's step into a stealthy near-future includes semi-matt black and liquid-metal titanium bodywork for minimal radar reflection, interchangeable side panels, different windshields for easy disguise, and a secret compartment with a sliding door.

BMW's future of urban mobility was revealed as part of their Vision Next 100 project in 2016: the Concept Link e-Scooter. It looks both futuristic and utterly practical, with enough groovy tech features to satisfy the urban mobility needs of the creative classes. Along with the possibility of total customization comes a wonderful maze of integrated technology: the instruments are gone, so important information is shown instead on a head-up display, while second-tier information goes to the touch-screen below the handlebar. And don't forget your biker's jacket at the casino table: the locked storage compartment can only be opened with a swipe over the right sleeve.

Acceleration
0–62 mph (0–100 km/h):
6.8 sec

Range
100 mi (161 km)

Power
Rated output 26 hp (19 kW);
max. 48 hp (35kW)
at 4,650 rpm

Torque
72 N m (53 ft-lb)

Top Speed
80 mph (129 km/h)

Battery
133 V air-cooled Li-ion
high-voltage battery

Weight
606 lb (275 kg)

Availability
One-off

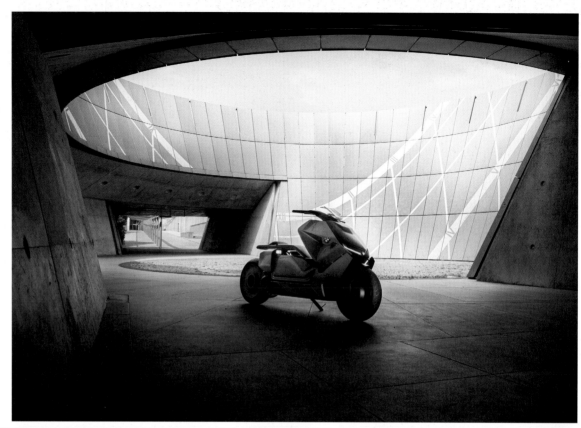

Acceleration
0–50 mph (0–80 km/h):
6.8 sec

Range
143 mi (230 km)

Power
28 hp (21 kW)

Torque
65 N m (47.94 ft-lb)

Top Speed
75 mph (121 km/h)

Weight
375 lb (170 kg)

Availability
Built on request

Shiny Hammer • HOPE

There's Hope for the Future

Shapes evoke memories, and designer Samuel Aguiar's e-bike called Hope stirs memories of the streamlined 1930s. It evokes fantastic designs like the Auto Union Type C racer, or the aluminum-clad planes built for the Schneider Trophy, with their romance of goggles and scarves. As designer Samuel Aguiar puts it, "'Hope' could have been designed a century ago or in a century." Aguiar worked for four years to develop this hypnotizing form, and explains, "I was looking for a design that you would want to hug, without any aggressive shape." The unusual two-tone bodywork is made of shiny aluminum plates in the center section, which are riveted to the matte black fiberglass bumpers (in front and back), then connected to a laser-cut steel frame. Beneath that shapely bodywork you will find the dependable Polish Vectrix VX-1 scooter serving as the base for this twenty-first century personal transporter. The Hope's flowing lines and gently pulsating surfaces are mesmerizing, suggesting that a daily commute can feel far from boring; it might even become an adventure.

Make Your Own Rules

Gloria Motorcycles is busting up the old motorcycle business model, because they think fighting big companies on their own terms is futile. They're rethinking the motor-cycle itself, framing it as a personalized statement and a high-tech mobility tool.

Gloria's ultra-simple layout is ripe for customization, allowing for an infinite variety of bodywork.

T

he electric motorcycle industry is not inherently a disruptor. According to the founders of Gloria Motorcycles, Antonin Guidicci and Benjamin Cochard, the existing model for selling electric cars—a combination of production, marketing, and distribution—has no future. Their radical approach is only possible with twenty-first century consumer culture, and, more significantly, they don't even think of their product as a motorcycle: it is a high-tech, fashionable mobility device. Their plan is to hack contemporary systems of communication (i.e., social media) to capture a demographic of urban youth who would never identify as motorcyclists. Gloria will invent a new segment of the marketplace by making a product the world didn't know it needed until they saw it on their phones, looking cool.

Gloria's Guidicci and Cochard are ignoring the usual R&D-based path to creating a motorcycle brand. Instead they are channeling their energies into harnessing social trends. "We think attacking the big players on technology, by following their own rules, is futureless," says Benjamin Cochard, who worked in the automotive financial industry before starting Jambon-Beurre Motorcycles with fashion photographer Antonin Guidicci. While Jambon-Buerre was a custom motorcycle shop for three years, building hip IC bikes, their last project,

a prototype of an electric flat-tracker with an outrageous 90 hp (67 kW), attracted so much attention that it switched on a light, inspiring them to change course and establish Gloria Motorcycles. They designed the architecture of a new e-motorcycle, but in the process they saw the need for a new kind of mass-production, one which would be possible only by using new

Gloria's founders come from fashion and finance – the perfect combination for a rule-breaking startup?

technology—but not necessarily motorcycle technology.

Gloria's disruptive path is not about building a network like Gogoro, or selling finely crafted, boutique machines to compete with their IC rivals. Their plan depends on three main elements: personalization, online ordering, and desire. With no Gloria dealerships, customers will order their bikes online, specifying the bodywork and colors they want, including a range of tanks, fenders, panels, and wraps. Customers can upload their own graphics, with the goal of making a unique, totally personal fashion statement. Gloria downplays the selling point of electric technology: they think what matters to a new segment of customers is not performance specs; instead they will be attracted to Gloria's easy ride and bold statement of identity. Cochard explains, "Our vision is 'do it differently.' When we decided to go for mass production, we analyzed what was going on, and the result was quite surprising. Big players are denying electric technology, just like Kodak when digital photography arrived." Given the rapid downfall of giants like Kodak and Polaroid, who 30 years →

"We believe the key to the future is about understanding that an electric motorcycle is not just a motorcycle anymore."

From a good idea to a way better one: Gloria shed its Jambon-Buerre cocoon to grab a chance at transforming an industry—not through tech, but social innovation, and a whole new relationship to wheels.

ago seemed invincible, their assessment is ominous.

Gloria also has a withering analysis of today's EV business: "For the moment, the EV segment is dominated by young start-ups doing a copy/paste of the existing industry business model, focused on developing technology. In this El Dorado rush, technology is not the key. Most EV startups focus on electrifying motorcycles. We believe the key to the future is about understanding that an electric motorcycle is not just a motorcycle anymore. It's an easy-to-use high tech product. So the existing distribution and communication system is totally obsolete, and can be reinvented almost from zero."

but also fashion objects capable of generating desire by expressing identity—seems a logical path forward. "First, we stop fighting on technical specs! Most people don't understand them anyway. Then we say goodbye to the whole distribution system: it made sense when vehicle maintenance required a qualified workforce, but with electricity, maintenance is more like an appliance workshop than a garage. And today, Amazon sells more appliance products than any brick-and-mortar distributor!"

The cultivation of desire has eluded the motorcycle industry in the twenty-first century, which has helplessly watched sales drop over time. Gloria's path is reminiscent of Honda's breakthrough into American homes in the early 1960s, when they marketed inexpensive, friendly, and stylish machines, glossed over tech, and won the world. Gloria similarly targets non-motorcyclists, "Especially young people and women that are completely forgotten in the current

Gloria isn't selling motorcycles: they offer high-tech mobility and a bold statement of identity.

system." Clearly, Gloria has the imagination to rethink an entire industry, and we see the appeal of both their product and their thinking: it is easy to imagine success for their fashionable transportation gadget. Gloria's analysis seems spot-on, but most of all, we love their attitude. ●

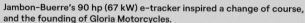

Jambon-Buerre's 90 hp (67 kW) e-tracker inspired a change of course, and the founding of Gloria Motorcycles.

Making the assumption that buyers no longer care about performance specs is both refreshing and good for the motorcycle industry. Bikes surpassed any normal human's ability to exploit their horsepower over 20 years ago, so the industry's arms race on power and handling has left potential new buyers cold. Gloria's take on motorcyles—that they are high-tech consumer products,

"Assuming buyers no longer care about performance specs is refreshing, realistic, and good for the motorcycle industry."

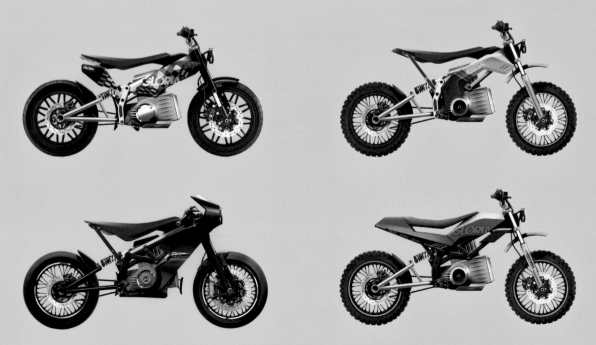

The thousand faces of Gloria. A motorcycle designed as a blank canvas for your personal style.

BMW x Krautmotors • C EVOLUTION "E-LISABAD"

The Origami Electragster

In the twenty-first century, could there be a more sensible choice for the responsible biker than an electric maxi-scooter? Probably not— but that doesn't mean you have to hide your true character. So, if you're into custom builds, this radical modification of a BMW C Evolution scooter from Krautmotors's Rolf Reick may tick all the right boxes. Considering the instantaneous delivery of electric power—electric engines produce peak torque at zero RPM—and the impressive acceleration this allows, it seems fitting to use it for ⅛ mile drag racing. The lowered front end and rigid rear give it a lower center of gravity perfect for this style of racing. The visibly raw powertrain oozes handmade character as does the little "starter battery pack" peeking out from behind the rear mini-fender. The stealth jet–like fairing may not be the latest trend in aerodynamics, but it certainly communicates "badass"— and that could prove very important for the acceptance of e-mobility in a wider audience.

Acceleration
0–62 mph (0–100 km/h): 6.8 sec

Range
100 mi (161 km)

Power
Rated output 26 hp (19 kW);
max. 48 hp (35kW) at 4,650 rpm

Torque
72 N m (53 ft-lb)

Top Speed
80 mph (129 km/h) (electronically governed)

Battery
133 V air-cooled Li-ion high-voltage battery

Weight
606 lb (275 kg)

Availability
One-off

Three Wheels to the Future!

The EV3 is "looking at the world of zero-emissions motoring from an entirely different perspective," claims the Morgan Motor Company's website. This perspective—besides sitting 8 in (20 cm) off the ground—is very refined: 1930s race cars meet 2020s electric mobility. Bespoke craftsmanship meets modern-world materials. The tubular steel chassis is covered by an ash wood frame and then clad with composite carbon panels. The wooden dashboard aims to please with its polished aluminum and brass details, its classic magneto-style switch for drive selection, and a digital screen. The typical excitement that comes form driving one of Morgan's three-wheelers will be further magnified by a 61 hp (46 kW) motor powering the sole rear wheel. And with their new technical partner, Frazer-Nash, developing the powertrain, the whole setup is sure to provide lots of fun for the gentlemen and gentlewomen driving it. Increasing the joy of anticipation could be opting for the special UK1909 Selfridges edition in collaboration with the world famous department store—and nine iconic british brands providing caps, gloves, goggles, and other suitable gear.

Acceleration
0–62 mph (0–100 km/h):
9 sec

Range
150 mi (241 km)

Power
61 hp (46 kW)

Top Speed
115 mph (185 km/h)

Battery
20 kWh Li-ion

Weight
1,102 lb (500 kg)

Availability
In production

kWh 2.6
Volt 51.8
Ah 50

kalk

Let Them Eat Kalk

**Meant for Getting Dirty, Swedish Brand CAKE's KALK
Off-Roader is Super Clean.**

S

tefan Ytterborn is a veteran of global corporations and indie manufacturing; he has worked with brands like IKEA and SAAB and he even started his own cycling and winter-sports protective-gear company (POC Sports), which he sold to Italian megabrand Dainese in 2015. POC Sports was an award-winner, used by athletes like Olympic skiing medalist Julia Mancuso and cycling champs Danny MacAskill and Martin Söderström. So it goes without saying that Ytterborn definitely understands the requirements of his clientele. For his new project, CAKE, he approached the manufacture of next-generation electric off-road cycles with total confidence, and the first product, the KALK, combines a seamless, modern aesthetic with high-performance underpinnings. The design team at CAKE took their cues from both downhill racing and enduro mountain bikes, combined with an MX motorcycle's ruggedness, to create an ultralight electric motorcycle weighing under 150 lb (72 kg). Everything about the KALK was designed from scratch—from the wheel hubs to the frame and bodywork—for lightness and strength. The frame and swingarm are built from aluminum, with CNC-machined joints welded together; they are incredibly durable, with minimal overall mass. The bodywork is made of featherweight carbon fiber, but it's the styling that sets KALK apart. The overall design succeeds by combining an ultraclean contemporary i-design vibe with a promise of off-road competence. It is an unusual mix, like an intelligent hammer by Jony Ive, and manages to look like the future while kicking up dust in the middle of a forest.

The KALK uses a 20 hp (15 kW) mid-mounted engine driving the rear wheel by chain, and is sourced from Europe, to CAKE's specification. The Öhlins suspension is fully →

What do you get when you cross a computer with a motocrosser? Something like this.

Cake's ultra-smooth, future-tech lines complement its underlying technology perfectly.

It's so clean it almost seems wrong to get it dirty, but those smooth shapes make cleaning easy.

"Light, silent, and clean electric off-road motor-bikes will make the era of noise, dis-turbance, pollution, and complexity a thing of the past."

Stefan Ytterborn

adjustable, and with three riding modes to choose from, riders of any age or capability can handle this ultralight, superquick machine. It can be ridden for two to three hours in a gentle manner, or for 45—60 minutes at full throttle on the track, where a rider can explore the top speed of +46 mph (+75 km/h). And the KALK isn't just lightweight: the quick, torque-y power-delivery of the electric motor makes for a serious yee-haw machine. The potential for wilderness fun is guilt-free; the KALK makes no noise, releases no toxic fumes, uses Trail Saver off-road tires that are kind to the earth, and has the possibility of going totally off-grid with CAKE's own solar-charging panels.

All of KALK's components are made in Europe, meaning the bike isn't cheap, but even the $14,000 (€12,000) price tag didn't deter customers: the limited edition of 50 machines sold out within six months. The KALK is a high-concept, limited-production design item, reminiscent of Philippe Starck's iconic Aprilia Moto 6.5, from the confident hand of a master. It is a design masterpiece with seam-less surface transitions and ultra-clean, pale color-tones, plus excellent suspension, brakes, and performance. The KALK manages the impossible by being, simultaneously, a serious dirt bike, a fun trail bike, and even more— a beautiful object. ●

Range
50 mi (80 km)
free riding 45—60 min

Power
20 hp (15 kW)

Torque
42 N m (30.98 ft-lb)

Top Speed
+46 mph (+75 km/h)

Battery
51.8 V (2.6 kw/h);
Li-ion battery

Weight
154 lb (70 kg)

Availability
In production

At the apex of twenty-first century design are products we intuitively understand as revolutionary.

The integration of motocross components with high-design chassis parts is seamless.

From the wheels to the frame to the sprocket to the forks, the KALK is a thing of beauty.

"We offer action and magic in combination with responsibility and the respect toward people and planet."

Stefan Ytterborn

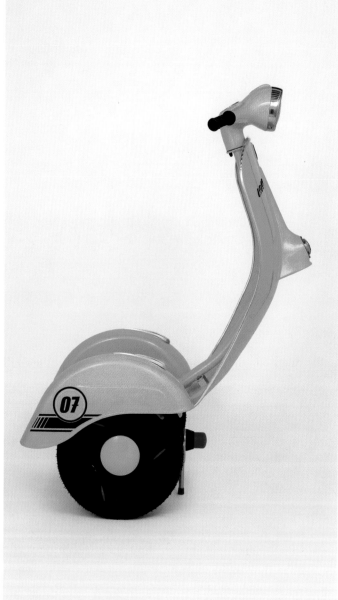

Range
18 mi (29 km)

Power
1.8 hp (1.35 kW)

Top Speed
12 mph (19 km/h)

Battery
55 V 450 Wh

Weight
55 lb (25 kg)

Availability
Ready to order

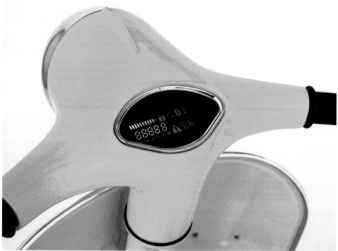

Bel & Bel • Z-SCOOTER

The Love Child of Vespa and Segway

At first glance, you might think this is just another cute helper-droid from the movies (think: WALL-E or BB8). But on closer inspection, you'll see that the Z-Scooter is not just another silly robot—it's a sexy, silly robot. Bel & Bel, known for their zany retro-futuristic designs, created the Z-Scooter as a conversion kit for the existing gyro-balanced Segway by NineBot. The result looks like a painfully geeky Segway had the best night of his life with a sleek Italian model from the 1950s. We say Italian because the Z-Scooter clearly takes its design cues from the iconic

retro scooter from Italy, the Vespa. And sure enough, the Bel & Bel designers credit their idea to Corradino D'Ascanio, the Italian aviation designer who turned his talents to creating the original Vespa scooter. While you might not be able zip around Rome with your date on the back of your Z-scooter (or you can try, but we don't recommend it), you can definitely use it to zip around Rome by yourself; with a range of 18 mi (29 km), this machine is a chic, green—if slightly quirky—solution to today's clogged and polluted urban transport systems.

A Faster Four-Seat Future

When a company with as strong a brand DNA as Porsche enters the electric car arena, you know it will be something sexy, and it will different from everything else. The Porsche Mission E is their first all-electric sports car, and that sports designation makes all the difference. While electric cars have been built since the 1890s, an electric sports car is another animal entirely: the key is not drag strip acceleration in the quarter-mile, it's about lap times on circuits like the Nordschleife at Nürburgring. So when you have a company with more than 30,000 racing victories under its belt, expect something

special. For an ultra-lightweight design for the Mission E, Porsche used a carbon-fiber monocoque that is so sturdy the car doesn't need a B-pillar. Additional carbon components carve off even more weight. With several Le Mans winners to serve as blueprints, you can expect them to get the aerodynamics right. Also, the battery-packed floor (which is recharged using fast inductive charging) is a big plus for handling, creating a very low center of gravity. The dashboard features five OLED round instruments, controlled by eye-tracking. This four-door, four-seater is truly electrifying.

Acceleration
0–62 mph (0–100 km/h):
< 3.5 sec

Range
310 mi (500 km)

Power
600 hp (440 kW)

Top Speed
155 mph (250 km/h)

Battery
90 kWh liquid-cooled
Li-ion battery, 800 V system
voltage

Recharge
15 min for 400 km range
(at 800 V)

Weight
4,409 lb (2,000 kg)

Availability
Concept car

Trefecta Mobility • DRT OFF-ROAD UNLIMITED & DRT SPEED PEDELEC

Range
62 mi (100 km)

Power
5.3 hp (4 kW)

Top Speed
28–43 mph (45–70 km/h)

Battery
30 V, 400 Wh Li-ion

Weight
84 lb (38 kg)

Special Feature
Wheelie mode

Availability
In production

Rocket Science

An international team of Dutch, Swiss, and German engineers with backgrounds in aerospace and automotive engineering joined forces to build a bike. And their high-flying experience shows: a foldable frame made from injection-molded aerospace-grade aluminum; strong and lightweight carbon-fiber wheels, interchangeable by quick-release mechanisms; an adaptable suspension fork whose stiffness settings can be electronically altered via the fly-by-wire handlebar controls or by the smartphone app. Your smartphone, by the way, sits in a waterproof phone dock for seamless mobile integration. The 5.3 hp (4 kW) drivetrain is hydraulically controlled and shifts automatically, while the interchangeable battery pack houses a 30 V lithium-ion battery good for covering over 62 mi (100 km) of mixed terrain. The DRT version is designed for off-road use, while the URB is the city bike, both available as a speed pedelec. This is what happens when engineers are not only allowed to dream but to build as well. And one last thing—there is an adjustable wheelie mode ... Yee-haw!

Savagely Independent

Build a bike. By hand. By yourself. That sounds like a lot of night shifts for a man who also works as the founder of an education start-up in New Orleans. But Matt Candler gives a very simple explanation for his mania: "I do this mostly because I want to make unique bikes that are fun to ride." He has succeeded in both by crossing a Suzuki with a Nissan, using a 2003 Suzuki Savage bike frame, mounted with a battery pack from a Nissan Leaf car, and driving it all through an EnerTrac wheel hub motor that can deliver 13 hp (9.7 kW) continuously and up to 40 hp (29.8 kW) at its peak. Then he threw in a bunch of component upgrades and a ton of passion. When you look past the fun stuff, Candler is quite serious about changing our habits: "We need to talk about getting off of oil sooner rather than later, and I think kick-ass, custom electric bikes can move the conversation forward." And who is more credible than someone who just goes and builds one of those kick-ass beasts himself?

Acceleration 0–50 mph (0–80 km/h): 6 sec	**Top Speed** 70 mph (133 km/h)
Range 100 mi (161 km)	**Battery** 2.5 kWh 96 V (LiMn2O4)
Power 40 hp (28.9 kW)	**Weight** 250 lb (113 kg)
Torque 100 N m (73.76 ft-lb)	**Availability** Custom

Night Shift Bikes • LEAFY SAVAGE

Inventing the Genre

The Bollinger B1 is the World's First Electric Sport Utility Truck.

W

hen Robert Bollinger wanted a truck for his farm—and wanted it electric—he discovered a tremendous void in the marketplace; nothing was available because nobody was making an electric off-road four-wheel drive vehicle. That was in 2014, and by June 2017 he'd built a team, designed a totally new type of vehicle, and shown the prototype Bollinger B1 to the world: the first "sport utility truck," a title that accurately encompasses the B1's various attributes of high power and speed, super-rugged construction, and impressive off-road capability, along with a wide versatility for potential uses.

Robert Bollinger studied industrial design at Carnegie Mellon University, and after a career in the New York City advertising business, set out in 2006 to create a line of organic hair and skincare products, which proved very successful. After selling that business, he started a grass-fed organic cattle ranch in the Catskill Mountains of upstate New York. A ranch needs trucks, which is how the Bollinger project got its start; how could there be no electric 4 × 4 trucks on the market? With this gap in the market, he saw the opportunity to create a totally new category of vehicle, which he dubbed the SUT.

After assembling a team of experienced designers in both vehicles and e-tech, Bollinger and automotive designer Ross Compton arrived at a shape that has the most distinctive and appealingly rugged design in the entire e-vehicle marketplace. Its boxy, flat-panel bodywork evokes the complete history of 4 × 4s, from the original Jeep to the latest Land Rover—future and retro all at once. But the only real nod to the past is found i n the instruments; round, black-faced, with chrome bezels. In its radical simplicity, the B1 is open to interpretation, although the ultimate takeaway →

The Bollinger makes a Humvee look over-complicated: its pure geometries are wildly rugged.

There's never been a personal utility vehicle capable of hauling a ship's mast in the cabin.

The rear seats fold away, or they can be removed, should you need to carry up to 72 sheets of ply.

"It goes against everything we think about electric vehicles... but that's on purpose."

Robert Bollinger

from its design is utility, reinforced by riveted panels and a total lack of "styling," making for refreshingly great design.

The B1's specifications are equally impressive. The frame (no expensive unibody here) is built from welded aluminum and houses two engines with a combined 360 hp (268 kW) and 640 N m (472 ft-lb) of torque, with a 0–62 mph (0–100 km/h) time of 4.5 seconds, and more ground clearance than any other 4 × 4 on the market—an actual ride-height of 15.5 in (39 cm). The hydropneumatic independent suspension can be pumped to a 20 in (51 cm) ride-height, is self-leveling, and is robust enough to carry the B1's weight in cargo—5,000 lb (2,268 kg). The rear seats fold away to carry a full 49 in (124 cm) wide load, and Bollinger claims it will carry 72 sheets of plywood! There's a tunnel opening in the dash-board for very long loads, making room for 24 12-foot-long 2 × 4s. Or, if the grille is open and the tailgate down, perhaps the mast of a sailing ship?

The Bollinger B1 taps into the fanta-sies of the young at heart everywhere. Its compelling, stark shape is an attention magnet; everyone wants to know about this vehicle—because Bollinger had the balls to make an ur-truck, the most basic thing possible, while still making it the 4 × 4 of the future. ●

Acceleration
0–62 mph (0–100 km/h,): 4.5 sec

Range
200 mi (322 kW)

Power
360 hp (268 kW)

Torque
640 N m (472 ft-lb)

Top Speed
127 mph (204 km/h)

Battery
120 kWh; Battery pack

Weight
3968 lbs (1800 kg)

Availability
Pre-production

The interior of the Bolinger is Spartan by design, as befits the world's first Sport Utility Truck.

With no motor under the hood, everything above the floor is carrying space, or storage.

The press and the public go nuts over the B1 because they got it right: keep it simple.

"It's a vintage sort of design language, but also embodies modern elements to keep it progressive."

Ross Compton,
designer

The Furious Non-Diesel

The mystery starts with the name: ETT Industries is an acronym for "escape through technology." They build and sell very special e-bikes, RAKER and TRAYSER, based on a patented monocoque frame with a sharp and futuristic look that's won them IF Design Awards. Originally founded in New Zealand in 2012, they are now a British company, based in London. The makers have a background in automotive and race car design (Founder, Jay Wen, studied in Milan and worked for brands like Peugeot and Citroën). The bike shown here is a one-off; the H1L was a 2017 project to create a limited-edition, collectable electric motorcycle produced in partnership with Universal Pictures for the *Fast and the Furious* film series. It features a unique design, combining ETT design language with TFF art direction. The stealthy look, laser-cut aluminum sheets, and the carbon-fiber seat ooze action-movie oomph. But if you're not the muscly guy in the black t-shirt—don't try this at home.

Range 62 mi (100 km)	**Recharge** 6 hrs
Power 8 hp (6 kW)	**Weight** 220 lb (100 kg)
Top Speed 81 mph (130 km/h)	**Availability** Limited edition (88)
Battery 72 V 60 Ah	

The Real E

Enzo Ferrari reportedly said that the Jaguar E-Type is "the most beautiful car in the world." These words became both an inspiration and an obligation for Jaguar Land Rover's new "Classic Works," which was opened in 2017, in Warwickshire, UK. When they took on the assignment of retrofitting the 1968 Series 1.5 Jaguar E-Type Roadster, the Classic Works decided to pull out all the stops; the result was the one-off Jaguar E-Type Zero. The car uses a zero-emissions, all-electric powertrain and can go from 0 to 62 mph (0–100 km/h) in just 5.5 seconds—one second quicker than the original! The most noticeable external changes are the modified instrumentation, the fascia panel, and the LED head-lights. Interestingly, the setup would fit any Jaguar with an XK engine (which covers most of the 1950s through to the 1970s), so this may be a test drive for a future branch of the business. As Tim Hannig, the director of Jaguar Land Rover Classic, phrased it: "Our aim with E-Type Zero is to future-proof classic car ownership." An honorable task, sir.

Acceleration
0–62 mph (0–100 km/h):
5.5 sec

Range
170 mi (270 km)

Power
400 hp (294 kW)

Torque
696 N m (513.34 ft-lb)

Top Speed
155 mph (249 km/h)

Battery
40 kWh, Li-ion 350 V

Weight
2,544 lb (1,154 kg)

Availability
Concept

2 × 2 Bike, True to Nature

Ubco of New Zealand has created a class of its own with their Utility Electric Vehicle (UEV)—a farmer's best friend and a thrill-seeker's dream. Their latest version, the 2 × 2, owes its name to its two in-wheel motors that deliver great traction, even on icy surfaces, essentially forming an all-wheel drive. Ubco says their electric moped "embodies a people-before-technology approach," meaning it is packed with tech meant to make life easier. The Electronic Control Unit (ECU) was developed from the ground up and controls the entire e-bike. Moreover, an app enables riders to switch between on- and off-road modes, or to enter into "Hunting Mode," which allows riders to turn off the headlight. The mechanical elements complete the multi-use vehicle with the iconic super-x frame, packed with strategically positioned attachment points so that users may mount anything however they want, be it a surfboard or a shovel. While the 2 × 2 is positively oriented toward off-road adventures, Ubco just received their road-legal status and is available for purchase in New Zealand, Australia, and the United States.

Range
75 mi (120 km)

Top Speed
30 mph (50 km/h)

Weight
139 lb (63 kg)

Power
2.7 hp (2 × 1 kW)

Battery
2.4 kWh

Availability
Production ready,
sold in NZ, AU, USA

Ubco Bikes • 2 × 2

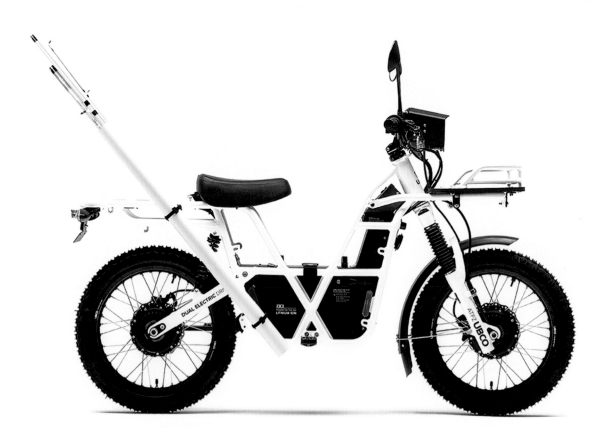

Cast-Forward a Century

Don't be fooled: Wannabe Choppers's latest build is anything but wannabe. The AlSi9Mg proves that craftsmanship and handmade parts will still have their place in the future of mobility, no matter how ecofriendly or futuristic the vehicles may become. The small company takes their design cues from 1920s and 1930s race bikes, but updates them with an electric motor. Barring the electric motor, everything about the AlSi9Mg is handmade, just like their inspiration. The bike is made entirely from hand-cast aluminum components made using century-old techniques such as a sand casting. In fact, self-taught motorcycle builder Enrico "Ricky" de Hass has been casting aluminum parts for himself since he was 11 years old. The particular alloy of aluminum used in the process is AlSi9Mg—hence the bike's peculiar name. With all this in mind, it comes as no surprise that it took Wannabe Choppers over 1,400 hours to build this motorcycle. At least the makers can rest assured it is exactly the e-chopper they imagined.

Range
75 mi (120 km)

Top Speed
62 mph (100 km/h)

Battery
52 V Li-ion

Weight
176 lb (80 kg)

Availability
One-off

Greetings from the Future

This Honda Urban EV Concept is an emissary from an adorable future. The exterior looks like a mix between a pug and a 1980s hatchback, but hiding behind its cute appearance are state-of-the-art features that you would expect to see on a Rolls-Royce and not a Honda (think: rear-hinged doors). But unlike a Rolls-Royce, the Urban EV is much smaller and much more, well, emotional: packed with sensors that are connected to an artificial intelligence system, the car can detect drivers' emotions based on their judgments.

It then makes recommendations based on the drivers' past activities. The plush interior resembles a 1970s living room pimped out with high tech: the chic wood-panel dashboard features a hightech wide-panel display, and cameras replace all the mirrors. On top of that, the Honda is ready for vehicle-to-grid applications. That means its battery can store energy and release it back into the home or the grid. And although this seems like the stuff of the future, this concept is poised to hit the market next year.

No One Is Ugly at Midnight

Apache Customs Shocks 1970s Nostalgics with an
Electric Endurance Racer.

F

or Apache Customs, character is the result of nurture, not nature. With a workshop on the idyllic shores of Italy's incredibly beautiful Lake Garda, one might expect Apache's founders, Giorgio Zambonia and Roberto Butturini, to produce two-wheeled Rivas. But even Shakespeare's "fair Verona" couldn't break them of their creative habit of building aesthetically challenging custom bikes. Their 2016 collaboration with Energica Motor Company was their first step into electric-bike turf, but you would never know it from looking at the Midnight Runner: the construction cleverly hides all clues to modernity. The outrageous rectangular headlamps mounted outside the dolphin fairing are a direct throwback to 1970s and 1980s endurance racing and are so distracting nobody notices that the period exhaust pipes are missing! It is a bold shape for the normally futurist, sport-bike-oriented Energica—and one they happily embraced to demonstrate that new tech can exist in any guise, even old ones. 1970s endurance racing was a particularly European sport, with its own design logic and a unique style that you either love or hate. This also sums up the press's reaction to the Midnight Runner when it was unveiled at the 2017 Verona Motorcycle Show: everyone loved to hate its style but had to admit it was the bike's very offensiveness that elevated it from bad to awesome.

Apache Custom Motorcycles was founded in Verona in 2015 and has established itself with punchy custom IC bikes with a punk vibe. Energica Motor Company of nearby Modena is a world leader in electric motorcycle manufacture, to the extent that is it the current supplier of racers for Dorna's electric MotoGP series, the Moto-E World Cup. Their Ego super-bike puts them on par with IC sport-bikes for performance, and their →

Retro is best when it's re-invented, mixing the best of today with the grooviest of yesterday.

The twin headlamps scream 'Endurance Racing', a wonderfully outrageous fashion statement.

With so many 1980s design cues to feast the eyes, nobody notices the lack of exhaust pipes!

"It's a race
in the
night,
a beacon
that lights
the way,
and it is
silence."

Giorgio Zamboni,
Apache Customs

Apache Customs • MIDNIGHT RUNNER

Eva streetfighter, with slightly less racy specs, was the foundation for the Midnight Runner. Apache chose to clad the Eva's contemporary chassis and battery pack with fiberglass bodywork that convincingly mimics Rickman-era café racers. It is a stylistic choice steeped in irony, seeing as 1970s endurance racing was an orgy of petro-gasmic noise and speed: the apex of up-yours fuel consumption and flamboyant petrochemical excess—the oil crisis be damned!—using the biggest, thirstiest motors that had ever raced on two wheels. Apache even chose a kind of iridescent british racing green paint scheme that is the exact color of motor oil spilled on asphalt, or the oozing oil slicks killing Bolivian forest groves. It is a brilliant mixed message, a statement of guilty cele-bration and adulation of a beloved historical style, with its daredevil swagger tamed by a clean-tech heart. The retro-future vibe is carried through with the LED lights hidden within those massive x-ray glasses posing as headlights, and a taillight panel reminiscent of contemporary Le Mans racers' rain lights. Only the digital instrument panel gives the game away, unless you are expecting an earsplitting open-pipe howl from the Midnight Runner as it streaks past at its 149 mph (240 km/h) electronically limited top speed. ●

Range
< 99 mi (< 160 km)

Battery
11.7 kW, Li-ion

Power
136 hp (100 kW)

Weight
569 lb (258 kg)

Torque
195 N m (143.82 ft-lb)

Availability
One-off

Top Speed
149 mph (240 km/h)

The Apache team is justifiably proud of their customized Energica Eva street racer: it rocks!

The big disconnect between e-moto and endurance racer is noise: the ear-splitting howl is gone.

Redefining Hyper

Coming from the birthplace of Nikola Tesla, Croatia's carmaker Rimac Automobili was practically destined to specialize in electric cars—hypercars to be exact. Rimac's Concept_One is all about power, from its nearly 1 megawatt battery to its four electric motors that, together, create a 1,224 hp (913 kW) drive. With that kind of power, even the experienced Richard Hammond, host of *Top Gear*, unintentionally crashed this million-dollar gem. If it was not the jaw-dropping specs, it was Rimac's sheer engineering ingenuity that convinced Porsche to acquire a stake in the start-up, hoping for technological spillovers. Rimac's core architecture is radically different: each of the two motors sits in the center of both axles, offering many advantages. Rimac had to keep all development in-house to achieve a high degree of integration between every component. This resulted in perfect weight distribution, an extremely low center of gravity, and most importantly, a supercar worthy of the twenty-first century—just as founder and CEO Mate Rimac set out to create. Consider the job done.

Acceleration
0–62 mph (0–100 km/h): 2.5 sec

Torque
1,600 N m (1,180 ft-lb)

Weight-to-power Ratio
1.55 kg/hp

Range
217 mi (350 km)

Top Speed
221 mph (355 km/h)

Availability
Limited production of 88 cars

Power
1,224 hp (913 kW)

Battery
Up to 90 kWh, LiNiMnCoO$_2$

Rimac Automobili • CONCEPT_ONE

The Dutch Fighter

VanMoof is where style meets substance. Their bikes win design awards and sell all over the world. But for them, good is never enough: here is a bike that recognizes its rider, defends itself against thieves, and features a near-invisible lock, activated by a kick. No need to carry heavy bike locks or fiddle around with chains. Instead, a tiny integrated module completely immobilizes the bike and auto-activates the theft defense. In case anyone tries to steal it, the bike responds with an earsplitting alarm and flashing lights. And if that does not work? The company has its own bike hunters who will track it down within two weeks or replace it. The coated aluminum frame comes in two versions, S2 and X2, the intelligent motor delivers smart support and turbo boosts for extra punch, and the upper tube features an LED matrix display for key information. You get encrypted Bluetooth and GSM connectivity, wireless firmware updates, and an app for everything else, including finding where you parked, if you are lucky enough to have this bike and a car.

Range
37–93 mi (60–150 km)

Power
0.33–0.66 hp (250 W–500 W) front-hub motor

Top Speed
20 mph (32 km/h) (US settings)

Battery
36 V 4 Ah

Weight
42 lb (19 kg)

Availability
In production

The Smarter Smartscooter

Gogoro has an incredibly popular e-scooter system that's already established in Taipei, Paris, Tokyo, and Berlin. They have important friends and big investors for their battery-swap system, but will win the urban mobility market through fun.

While the rest of the world slept, China experienced a silent revolution. A change in their laws in 1999 allowed e-scooters to be classified as bicycles (with no license required) if they met a 20/40 rule: maximum speed 20 km/h (12 mph), maximum weight 40 kg (88 lb). In the mid-2000s they followed up with a total ban on IC scooters in big cities, and consequently the Chinese e-scooter industry blossomed, seemingly overnight. The industry was suddenly bursting with dozens of manufacturers who lead the world in EV production, having sold an estimated 200 million e-scooters to date. Although it went unnoticed by the media in the West, two strokes of the legislative pen had instantly transformed mobility for a billion people. As the EV industry in the rest of the world struggles to find its legs in the marketplace, the Chinese industry is simply enormous, and virtually unknown elsewhere ... unless you visit Shanghai, or New York City, where pedestrians dodge mad delivery riders on the sidewalk carrying yet another Seamless dinner order on a Chinese e-bike or e-scooter. They are incredibly popular because they are cheap, and they fill an important niche.

The e-scooter eruption in China was a mass-market response to an exploitable void, but creating a niche in markets dominated by IC vehicles— →

Luke Horace has a better idea: replace every gas station with a battery-swapping kiosk.

and no legislation in sight to ban them—takes a visionary.

One such man with a big idea is Taiwan's Luke Horace, who is challenging the very concept of the gas station. His Gogoro Smartscooters are hugely popular in Taiwan, with 85,000 customers using a system of "stations" where riders simply swap battery modules and ride away. The process takes seconds, costs one-third the price of petroleum, and has spread to pilot programs in Berlin and Paris (in partnership with Bosch) with the COUP app-based ride system. Horace explains, "Electric must be adopted on a mass scale, not just for the 1% of the population but for all people living in densely populated cities. People gravitate to Gogoro's design, performance, and connected innovation, and our customers in Taiwan have ridden nearly 300 million kilometers (186 million miles): they swap batteries 45,000 times a day!"

The design of the Gogoro is a major factor in its success, with a smooth, rounded, friendly-futuristic shape, with exceptionally easy access to the batteries and mechanicals. It is an inherently appealing design, and riding one is incredibly easy as is keeping them charged: in Taipei, there's a charging station every kilometer, and the European projects are building up to a similar saturation.

Good design wins the day, and Gogoro's friendly-egg curves are an invitation to ride. 85,000 customers in Taipei have already given their thumbs-up to stylish and easy urban transportation.

nlike the ubiquitous cheap Chinese e-scooter, the Gogoro has real performance to back up its looks, with a maximum speed of 55 mph (89 km/h) for the S2 model, making it a 125 cc e-quivalent, rather than the 50 cc rating of most commuter e-scooters. The concept, design, execution, and delivery of the Gogoro system ranks it with Tesla for industry disruptor bravado, but while Tesla brought sexy to electric, Gogoro grabs customers by their fun bone. That fun factor is the crack in the IC edifice that will eventually make the wall crumble; e-scooters are poised to be the first e-vehicle to the break through to mass adoption. "For Gogoro, we wanted to bring a new energy model to the industry and needed a vehicle that could showcase battery swapping while having an amazing design and function, incredible performance, and a level of connected sophistication that had never been seen." They've scored on all points.

Major banks and environmentalists like Al Gore are standing behind Gogoro, as what is good for this company is good for the world. Luke Horace calculates that "Gogoro customers have so far saved more than 14,428,265 L (38,115,440 gal) of gasoline and nearly 25 million kg (55 million lb) of CO_2, the approximate amount of CO_2 that 2 million trees consume every year, and 98% of the material by weight in the batteries can be repurposed. We want to provide a sustainable path away from gas, to everyone in cities." If you've ever walked the summer streets of Paris with eyes watering in the two-stroke moped haze from the stoplight drag races, you understand the electric future cannot come fast enough. ●

Swapping batteries takes seconds, requires no tools, and your hands stay clean. So easy.

"Gogoro customers saved 14,428,265 liters of gasoline and 25 million kilograms of CO_2, the approximate amount of CO_2 that 2 million trees consume every year."

Gogoro's network of battery stations eliminates the need to plug in batteries at home, or stretch a cord to your scooter. It's clean and easy, and might overcome the single biggest hurdle to e-scooter adoption.

The Electric Gauntlet

"Athlete and gentleman" in one package is a much sought-after combination, and not just when using a dating app. Clearly there is a market for an electric bike that is good for both flat track racing and urban commuting; the Alta Redshift ST Street Tracker Concept is a conceptual look at what street tracker fans might expect from the future. The combination of 51.5 N m (38 ft-lb) of torque and a low 250 lb (113 kg) weight suggests a good time in the concrete jungle—"The Future of Fast", as they call it at Alta.

The selfdeveloped battery pack crams 5.8 kWh into 67.9 lb (30.8 kg) with a maximum of 350 V. That is a lot of power from a small, energy-dense, waterproof, and durable pack. It is not surprising that Harley-Davidson bought shares in the company to keep pace with the times, developing competitive electric bikes. As Derek Dorresteyn (cofounder and chief technology officer) puts it: "We try to be kind, but our bike and customers are in effect telling the industry, 'your product is obsolete.'"

Acceleration	Power	Top Speed	Weight
0–60 mph (0–97 km/h): 3.5 sec	42 hp (31 kW)	80 mph (129 km/h)	250 lb (113 kg)
Range	**Torque**	**Battery**	**Availability**
60 mi (97 km)	51.5 N m (38 ft-lb)	5.8 kWh, Waterproof Li-ion 350 V	Concept

Alta Motors • REDSHIFT ST STREET TRACKER CONCEPT

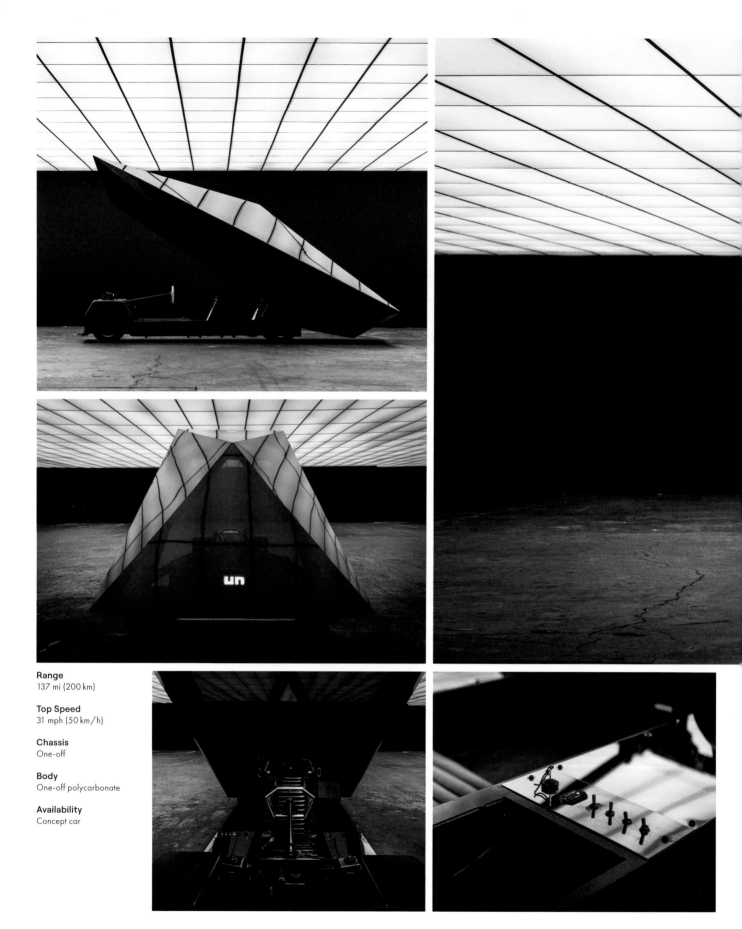

Range
137 mi (200 km)

Top Speed
31 mph (50 km/h)

Chassis
One-off

Body
One-off polycarbonate

Availability
Concept car

The Geometric Ghost of Gandini

Want to blow up the Internet? Build something spectacular. In this case, that took distilling the essence of a legendary Italian design: the Lamborghini Countach, designed by Marcello Gandini and first shown in 1971. The Countach was extremely futuristic, using radically minimal, faceted bodywork that became Gandini's signature. But if you squint at the Countach, or render it at super low resolution, what you might see is United Nude's Lo Res Car. Built as a design experiment, it was conceived in 2016 as part of the Lo Res Project of Dutch footwear brand United Nude. The transparent "smoked" polycarbonate body has no doors—the entire top hinges up from the back like a futuristic oyster. There's a sharp hexagonal steering wheel and an ultra-minimal interior that feels like a hard-edged and dangerous future, but it still tickles the same brain synapses as Gandini's wonderfully exalted prototypes from the 1970s. That was a carefree time, when designers felt free to experiment. Apparently, some of them still do.

Mopeds and Electric Freedom

Whole generations of youngsters got their first taste of freedom on a moped, and Onyx, based in San Francisco, wants to bring that tradition back. Taking their design cues from the vintage mopeds of 1970s Japan and Austria and adding electric motors, they are bringing mopeds into the modern age. Their models retain a raw charm thanks to industrial designer and Onyx founder Timothy Seward. Having worked for Nike, Samsung, and Google, his strong sense for emotive visuals extends to his vehicle design. At Onyx they build tough, using steel frames and series moped parts attached. Except you will not find a two-stroke engine and its trail of smoke. Instead, an electric motor delivers 750 watts, fueling Onyx's mission to bring back moped culture. The RCR and CTY model, comparable to 125 cc and 50 cc, strike a balance between fun and style. For the stronger RCR, this means impressive speeds up to 60 mph (97 km/h) with a range of 75 mi (121 km). For the CTY, a turn of the throttle accelerates the e-moped to over 30 mph (48 km/h) for any trip within a 40 mi (64 km) range.

Acceleration
RCR 0–30 mph
(0–48 km/h): 3.0 sec
CTY 0–15 mph
(0–24 km/h): 5.4 sec

Power
RCR 7.2 hp (5.4 kW)
CTY 3.3 hp (2.5 kW)

Battery
RCR 72 V
CTY 48 V

Torque
RCR 182 N m (134.2 ft-lb)
CTY 148 N m (108.2 ft-lb)

Weight
RCR 120 lb (54.4 kg)
CTY 85 lb (38.5 kg)

Range
RCR 75 mi (121 km)
CTY 40 mi (64 km)

Top Speed
RCR 60 mph (97 km/h)
CTY 30 mph (48 km/h)

Availability
Pre-production

Onyx Motorbikes • RCR AND CTY

Ride Like a Champion

Once upon a time, "a few lucky folks got their thrills by riding motorcycles around large oval wooden-board tracks" That's how e-bike maker Vintage Electric tells the story. Inspired by this old pastime, the company started in 2013 in Santa Clara, California, with a single product: the Tracker. Sporting the look and feel of an early Board Track racer combined with the casual, suntanned appearance of a beach cruiser; this is an everyday bike that is primed for fun on weekends. Tracker and Tracker S share the same chassis and specs except for the suspension fork and higher-capacity battery on the S. The battery is enclosed in VE's signature aluminum battery box. Both feature five levels of pedal assist, so the hub-mounted motor can power your adventure at speeds of up to 36 mph (48 km/h). The hydraulic-disc brakes offer control while the regenerative rear brake returns power to the battery. The rugged, hydroformed aluminum frame is fitted with a traditional cone-shaped metal headlamp (LED, nowadays). Get ready for the Motordrome, buddy!

Range
50 mi (80 km) / Scrambler
75 mi (121 km) / Scrambler S

Power
93 hp (700 W) /
1.48 hp (1.1 kW)

Top Speed
36 mph (48 km/h)
(optional race mode;
for use on private property only)

Battery
52 V 12.5 Ah LiNiMaCo

Availability
In production

Radical Innovation in a Friendly Shape

While its body resembles a cross between an insect and a Junkers Ju 52, the Johammer J1 is one of the most radical and well-thought-out e-motos on the market. It is also the first in its category to achieve a range of 124 mi (200 km). The chassis incorporates surprisingly original ideas, with a horizontal aluminum-beam frame connecting the front and rear mono-shock absorbers and the hub-center steering up front—a combination that eliminates the need for a traditional frame. The unusual, shapely bodywork is cast from corrugated 0.28 in (7 mm) thick polypropylene, with a ribbed surface reminiscent of a vintage Junkers aircraft fuselage or a Citroën H van. Its rounded shapes echo the unusual chassis design and remind us that traditional motorcycle design—a rider straddling a fuel tank—is no longer needed, proving the point that new shapes are not only possible for EVs, they are necessary. The J1 needs no external instruments, because the hi-tech rear-view mirrors display all relevant information. The battery pack was developed specifically for the J1 by Nordfels, and once it has lost its peak performance—after 62,100 mi (100,000 km)—it will start its new life, storing electricity in solar power plants.

Range
124 mi (200 km)

Power
15–21 hp (11–16 kW) max.

Torque
220 N m (162.26 ft-lb)

Top Speed
75 mph (120 km/h)

Battery
72 V 12 kWh Li-ion, air-cooled

Weight
392 lb (178 kg)

Availability
In production

Johammer • J1

Night Ride

"We are pioneers. We know where the world is headed. Sometimes it gets lonely out in front, so let's do an electric night ride."

Like the first years of the automobile, early adopters of EVs are pioneers. They know where the world is headed, but might have only seen one or two other e-riders on the streets; they'll stop and chat, or just wave, because it is a small club so far. Every rider is an ambassador too, proof that some are brave enough to shut out the noise and naysay and simply enjoy the ride, adapting to something a little different. But not that different: e-everything is awfully convenient. Still, EV adoption is an uphill battle against a century-old industry, even if nearly everyone understands the damage that industry has done. So e-riders, all of them, remain pioneers, and as bold and isolated as they might feel, hooking up with like-minded folk bolsters their fighting spirit. Enter the Electric Night Ride.

A non-commercial, social ride that assembles e-riders in towns across Europe. The Ride is independent of brands and

sponsors, and simply wants to bring passionate e-riders together for fun, and to raise visibility of the EV movement. It unites regular EV users, future users, and interested citizens who get to see the huge variety of e-wheels available today. A silent ride at night seemed the perfect combination, and the organizers say "We equally aim at visibility, before, during, and after the rides to show the public which mobile options there are and how much fun they are, either through the event itself or through press coverage."

The ad-hoc world record for an all-brands EV tour was 69 riders in Paris in 2018, up from 43 in Lucerne in 2017, and 55 in the inaugural ride in Antwerp in 2016. The next event will be held in Stuttgart in 2019, with more to follow. Keep your eyes peeled on social media, or start an Electric Night Ride in your town! Let's save the world, one Night Ride at a time.

"Does it use a battery for power? Bring it.
The revolution will be silent,
except for the music of our laughter."

Lingyun Technology · 1703 Prototype

Meet the Jet's Son

Range
< 62 mi (<100 km)

Top Speed
62 mph (100 km/h)

Availability
Prototype

While automotive concepts are seldom built in the real world, sometimes a radical vision is held in a designer's mind, changing shape and blossoming years later. Take the Gyron, a Ford concept car from 1961 that used gyroscopes to stay upright, resembling something from a futuristic TV cartoon. It never went into production, but fascinated and inspired Zhu Lingyun, a graduate of China's Beihang University, a school known for its aeronautics and astronautics research. He theorized that a self-balancing single-seater, single-track vehicle would save a lot of space on the road and conserve energy in motion. If built to be safe and reliable, it could be a part of the urban mobility solution for the future, and a lot of people seem to agree; he founded a company in 2014 with angel investments, then had a successful first round of financing with international investors; the company is currently valued at $60 million (€51 million), according to Zhu. "The gyrocar carries people's imagination about future transportation. I have to make it," says the founder. And maybe that is what our future needs: a little less conversation, and a little more action, please.

The Modular Café Racer

As naked as a bike can be, the DCH1 is the brainchild and personal project of Spanish mechanical engineer and industrial designer Pablo González de Chaves. The bike is handbuilt from scratch, using only a handful of factory-made parts—brake calipers, lights, rims, forks, and shock absorber. The chassis consists of tubular steel structures bolted to aluminum CNC-machined parts. The battery pack, contained in an aluminum housing, also serves a structural function, by reinforcing the frame. The aluminum swingarm is composed of extruded profiles welded to CNC-machined parts, and the chassis design is completely modular, allowing for easy adjustments to the bike's geometry. This industrial style also defines the aesthetics: minimalist and mechanical, it wavers somewhere between a futuristic gadget out of a superhero's garage and a self-built version of a café racer. In this respect, the DCH1 represents a future of mobility that is personal and unique, achieved through skilled craftsmanship. Good to know that geniuses are still among us.

Range
40 mi (64 km)

Power
Rated output 14 hp (10 kW);
max 41 hp (30 kW)

Torque
100 N m (73.76 ft-lb)

Top Speed
65 mph (105 km/h)

Battery
76.8 V bespoke battery pack

Weight
253 lb (115 kg)

Availability
Prototype

Dechaves Motion • DCH1 ELECTRIC NAKED

The Day After Tomorrow

Lamborghini designer Mitja Borkert has a recurring problem: how to reimagine the future in radical terms—again. His latest project tackles the problem by focusing on five areas: energy storage systems, innovative materials, propulsion systems, visionary design, and emotion. Developed in collaboration with MIT, Lamborghini's hyperfuturistic concept car, the Terzo Millennio, envisions supercapacitors, a next-generation battery technology that can rapidly store, and release, huge amounts of energy—crucial for blistering acceleration. With four separate motors, one in each wheel, designers had maximum freedom, which they used to create a chassis with ultrabold lines. Since the body is made of highly conductive carbon fiber, it can store extra energy and will be able to, in theory, heal itself via micro tubes that channel healing chemicals to cracks in the carbon structure. For added saftely, the Piloted Driving simulation can guide a driver around a racetrack before allowing the driver to take over the wheel and drive into the third millennium.

Because You Need Tunes

Electric mobility is charged with purpose: it might help to solve humanity's biggest problems, from global warming to overcrowded cities. But besides the voice of reason, there is also "the joy in journey," as they say at the Slovenian manufacturer Noordung, which plans to return joy to the rider using music. The most obvious difference between this café racer-style e-bike and all others is the detachable boombox; both a battery housing and personal stereo in one, it plays music from your mobile phone through two speakers, assists with your pedaling, and charges your personal devices, all while analyzing air quality on your route. This miniature power plant sits on top of a carbon-fiber frame, accentuated by a leather saddle and grips. Speaking of innovative ideas, Herman Potočnik, nicknamed "Noordung," was a Slovenian rocket engineer and astropioneer in the early twentieth century who developed a concept for a space station that later inspired Stanley Kubrick during the making of *2001: A Space Odyssey*. They aim high at Noordung.

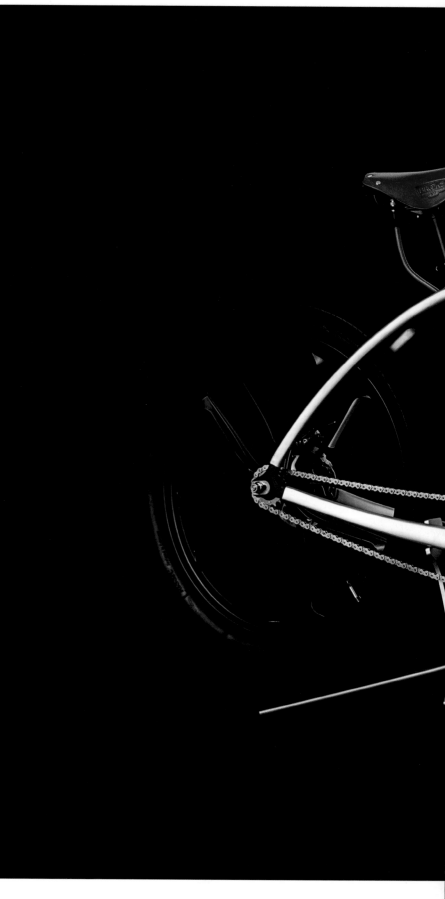

Range
18 mi (30 km)

Power
4.75 hp (200 W)

Battery
400 Wh Li-ion, 30 V

Weight
34 lb (15.4 kg)
Boombox 6 lb (2.7 kg)

Availability
Limited production

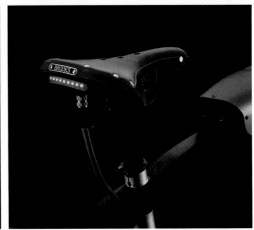

Noordung • NOORDUNG ONE "ECLIPSE & MOONLIGHT EDITION"

Kids, Play Quietly!

Off-road is where riding electric really makes a difference: no noise, no fumes, no grease or leaking oil. But making a difference can still be big fun, especially if you trust in a company that claims to deliver bikes "ready to race." KTM isn't new to the e-bike business—their development began around 2008, and it shows: the dust- and waterproof brushless electric motor provides 22 hp (18 kW) of peak power, and the water-cooled control unit delivers reliable function under any conditions. The 3.9 kWh battery pack can be swapped for a fully charged one or be charged while mounted, and with economy mode selected, the E-XC recuperates deceleration energy during coasting and braking. The lightweight steel-aluminum composite frame ensures rigidity and stability even in high jumps, and a high-strength polymer subframe at the rear reduces overall weight to 243 lb (110 kg). Suspension, wheels, brakes, and other equipment, ranging from footpegs to mudgrips, are as professional as you would expect from Europe's biggest motorcycle manufacturer. This may not be the most futuristic challenger, but it could be a wise choice, depending on the battleground. And this is just the beginning: according to KTM's CEO, Stefan Pierer, the company's transition to electric motorcycles is in full effect. Coming up next: the E-SX-Mini—for real kids.

Range
1.5 hours off-road –
48 mi (77 km)

Power
Peak 22 hp (18 kW)
Continuous 15 hp (9 kW)

Torque
42 N m (30.98 ft-lb)

Top Speed
43 mph (70 km/h)

Battery
3.9 kWh Li-ion KTM
Powerpack

Recharge
in 80 minutes

Weight
243 lb (110 kg)

Availability
In production

The Geometrist:
Undoing Expectations

Joey Ruiter's Radical Shapes Throw History and Design Conventions out the Window.

S

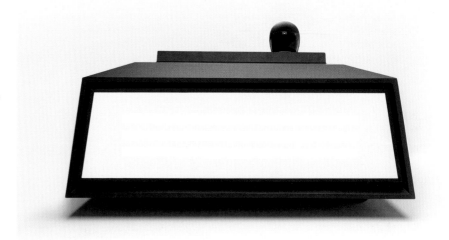

ci-fi movies are among the biggest-grossing films in history, and in 2017, electronica supplanted hip-hop as the biggest-selling music genre worldwide. Clearly, we love the futuristic, the synthetic, and the ultra-modern. Why then do we still drive boring sedans and motorcycles that could have been built in 1980, or even 1940? The mobility industry lost its taste for fantasy in the 1970s, perhaps because the Oil Crisis exposed the ugly geopolitical assumptions behind The Future, as imagined in the West. While some designers have explored relentless geometries in neo-wedge supercars, industrial designer Joey Ruiter has invented a new modernity, using radical new shapes for motorcycles, cars, snowmobiles and boats (and even furniture). In the process, he is challenging design hegemony with his outrageously simple vehicles.

The Moto Undone prototype of 2011 is the anti-motorcycle, discarding everything motorcyclists love about two wheels—cool engines, erotic shapes, nostalgic cues—in favor of an absolutely pure idea. It exists in our world like Stanley Kubrick's monolith in *2001: A Space Odyssey*, agitating knuckle-draggers to throw their bones because, to them, its shape is impossible. Moto Undone's mirrored flat planes exist on no other motorcycle because they are absurdly simple, and because its unyielding surfaces concede nothing to humans—or nearly nothing: the only clues for a rider are a beveled seat carved from the rectangular cuboid body, and barely protruding pegs for the rider's hands and feet. An important hint to functionality is the lightning-bolt gap across the front third of the body, creating the possibility for turns and front suspension; this makes Moto Undone more than just a thought experiment: the possibility of owning such an object is mind-blowing.   →

What is the shape of the future? Or, if you felt free to build anything, what shape would it be?

Joey Ruiter's levitating black trapezoid is disturbing because everything about it is wrong.

With no human-scale clues barring seats and a steering wheel, it could be an alien artifact or sci-fi prop.

The Consumer Car cuts the umbilical cord to the horse-drawn carriage, and the living room sofa.

"I wonder
why things
need to be
the way
they are,
and if
they don't,
then what
else could
they be?"

Joey Ruiter

J.Ruiter • CONSUMER

Turning his attention to four wheels, Ruiter's concept car, which he calls Consumer Car, is an electric automobile like no other. Its spare, trapezoidal shape is still clearly a car, with seating for four and a steering wheel, but that's about it for details. The only concessions to its occupants are footwells for clambering into its doorless body, a little padding on the bench seats, and the most abbreviated windscreen ever. No instruments, no handles, no top, no headrests or rollbars, and—at least from the outside—no wheels, as they are totally hidden. The Consumer seems to float over the road. Its mirrored grille emphasizes otherworldliness with a rebuff to curiosity, while the relentlessly all-black-everything finishes absorb all light, and the three LED headlight strips that shine through the grille are as alien as off-world landing lights. Ruiter is the first vehicle designer to grapple with the dystopic, human-unfriendly future imagined in twenty-first century cinema, and his work succeeds not by making fantasy props, but by creating actual near-future production prototypes. The future they suggest is hardly chic or sexy; instead, it is dominated by artificially intelligent overlords unconcerned with human needs but capable of building fascinatingly pure machines. ●

Acceleration
(0—30 mph, 0—48 km/h,):
3 sec

Battery
1,000 W/ 48 V

Range
30 mi (48 km)

Weight
150 lb (68 kg)

Power
1.34 hp (1 kW)

Availability
Prototype

Top Speed
30 mph (50 km/h)

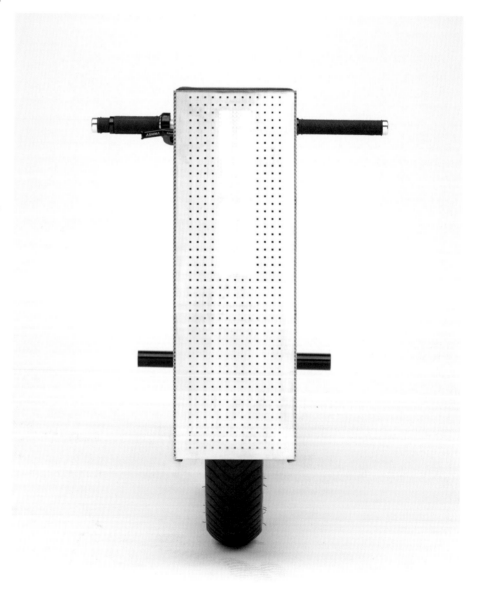

The Moto Undone is exactly that: the negation of every motorcycle ever built, before this one.

Motos need humans to move, but no "centaurs" will merge with this machine: it refuses the organic.

"What
disrupts
what
we expect
to see?
And helps
us look
beyond
what
we already
know?"

Joey Ruiter

Your 3-wheeled Soulmate

Half car, half motorcycle, this three-wheeled two-seater promises to deliver maximum smiles per mile. Vanderhall's first electric vehicle, the Edison^2, is a big step into the future of personal transportation: small, lightweight, and—most importantly—fun to drive. It uses retro design elements to evoke an old-timey roadster feel while also taking design cues from Formula 1 monoposto racing cars from the 1960s. It uses a tadpole configuration with the two front wheels in charge of both the drive and steering, while the sole rear wheel remains passive. The dual electric front motors deliver the equivalent of 180 hp (134 kW)—

more than enough when seated only a few centimeters above ground with no roof over your head, inside a very light vehicle with road-hugging geometry. The Edison^2 is hand-built and constructed around an aluminum platform, then clad with plastic composite bodywork. It also features a lot of car-like amenities: ABS, traction control, brake assist, steering assist, a roll bar, and even heated seats. The net effect is not merely transportation from A to B; this is about revitalizing your "driving soul," as the makers call it, making it more of a leisure sporting vehicle than a humdrum commuter. Niche, but nice.

Acceleration
0–62 mph (0–100 km/h):
4.0 sec

Range
200 mi (322 km)

Power
180 hp (134 kW)

Torque
250 N m (184 ft-lb)

Top Speed
105 mph (169 km/h)

Battery
30 kW Li-ion

Weight
1,400 lb (635 kg)

Availability
In production

Vanderhall • EDISON²

Power Hidden in Wood and Steel

Avionics's story begins on a Baltic beach where founders Jaromir and Bartek first met and quickly discovered their shared taste for adventure, speed, and the wonders of nature. Not long after, they began work on a machine that would combine all those elements. The result was the V1, a pedelec bike made entirely without plastic parts. The throttle levers, handlebar horns, and even the front lamp housing are all carved out of jatoba wood, which is precious, durable, and extremely hard-to-work. The same material was used to shape the bike's core: an elegantly carved wooden case protecting the electronics and battery. Leather belts strap the box to the openwork steel frame. The upper pipes are shaped like a wing and hold a shock-absorbing saddle—much needed suspension when sitting atop a 6.7 hp (5 kW) electric motor. The street-legal versions are tamer, ranging from 0.33 to 1 hp (250 to 750 W). But even at its lower speeds, Avionics's V1 is designed with flight in mind, sure to make you feel like you are soaring even while you get down and dirty, speeding along a dirt trail.

Range
75 mi (120 km)

Power
6.7 hp (5 kW) max.

Torque
125 N m (92.2 ft-lb)

Top Speed
36 mph (58 km/h)

Battery
1,175 kWh Li-ion

Weight
110 lb (50 kg)

Availability
Pre-orders open

Bouldering on Two Wheels

Electric racing vehicles are kicking butt in hill climbs, road racing, dry-lakes racing, motocross, etc., and another super-specialized off-road sport may soon witness an evolution: off-road trials. Trials riding is a balletic mix between parkour and motorcycling, where extremely skilled riders jump from boulder to boulder, pivot atop impossibly narrow logs, and climb vertical obstacles that seem simply impossible. Yamaha's first e-trials prototype, the TY-E, debuted in 2018, and began winning immediately under the skilled hands of Kenichi Kuroyama, a 39-year-old, 11-time Japanese trials champion. Trials motor-cycles are extremely light but powerful, and an electric motor's capacity for 100% torque at 0 rpm is ideal for leaping up from a standstill. That said, power control is still the main issue. Oddly enough, Yamaha found that a mechani-cal clutch gave smoother starts, while an engine flywheel gave better traction; both were necessary to deliver exactly the right kind of power to win a trials competition. The TY-E's body is all carbon fiber, with a first-ever use of nano-film technology for the bodywork to save a few grams. The TY-E won from its very first competition, and it appears another sport will soon go quiet.

Yamaha • TY-E

To Infiniti, and Beyond!

The Prototype 9 Mixes Aerofuturism with a Retro Grand Prix Vibe.

W

hile all the vehicles in this book started as a sketch, some renderings have the mojo to get an entire industrial team behind them. The sketches from Infiniti's senior vice president of global design, Alfonso Albaisa, were so striking they inspired several departments at the Infiniti factory to collaborate on making the Prototype 9 a reality. His renderings for an open-wheeled racer mused on the early history of Nissan, with its origins in the Prince sports racing car, the first true racing car designed by a Japanese manufacturer. The Prince R380 was built in 1965 as a purpose-built Le Mans-style racer, meant to compete with cars like the Porsche 904. It used a tubular Brabham chassis, a tuned Prince Skyline six-cylinder engine, and beautiful aluminum bodywork built by the team at the Prince factory. The Prince Skyline became Nissan's premier sedan (and a Japanese domestic market legend) when Nissan bought Prince in 1966, just after the R380 defeated all the competition in the Japanese GP that year. Reflecting on that successful effort by the team at Prince (which is part of the Nissan family), Albaisa wondered what a 1930s or 1940s Japanese sports-racer would look like.

The sketch of a "lost" Nissan mono-posto racer was intended as a thought experiment, but when Albaisa shared it with his colleagues, the idea proved so compelling to Infiniti's design team they conspired to have it built. At their design studio in Atsugi, Japan, each of Infiniti's specialist component designers—for seating, instrumentation, bodywork, and even cutting-edge EV technology—made a contribution to building the car. Meanwhile, the project was kept quiet at the factory, even as a full-scale mock-up was built in clay! When the model was complete, the physical production team felt compelled to turn the model into a fully →

"Prototype 9 has been a labor of love for many of us."

Alfonso Albaisa,
Senior Vice President,
Global Design

An engine-turned dashboard and all-aluminum shifter assembly are notes from historic racers.

Infiniti • PROTOTYPE 9

functioning car. The project was then moved to a quiet section of the Nissan factory near Yokohama where the car, an after-hours passion project that had turned into a full-scale concept car, could be built in seclusion. A team of *Takumi* (Nissan's master fabricators) built a tubular ladder frame in the style of vintage racers and hand-hammered sheet steel to create the aerodynamic retro-futuristic bodywork, including the fabrication of complicated grill-work, vents, and knock-off wheel hubs. The cockpit was upholstered in premium black leather, and the simplified instrumentation prevents driver distraction, although the dashboard is finished in a beautiful engine-turned aluminum, as was common in the 1930s but is practically a lost art today. The skinny wire wheels and cross-ply tires confirm the retro intentions of the Prototype 9, but the powerplant and sleek bodywork all speak to the future and were designed by the Nissan Advanced Powertrain Department as a prototype motor system. The Prototype 9 is completely drivable, with 148 hp (120 kW) and (more importantly) 320 N m (236 ft-lb) of torque, allowing for a top speed of 106 mph (170 km/h), and enough battery for a 20-minute flat-out run on the racetrack. ●

Acceleration
0–62 mph (0–100 km/h):
5.5 sec

Range
20 minutes full-throttle track time

Power
148 hp (120 kW)

Torque
320 N m (236 ft-lb)

Top Speed
106 mph (170 km/h)

Battery
30 kWh

Weight
1,962 lb (890 kg)

Availability
Prototype

At certain angles, the Prototype 9 is a pre-War monoposto Grand Prix racer: a Silver Arrow.

While sketched out on a computer, the clay mockup, and actual bodywork, were all built by hand by Takumi.

Despite the never-was retro configuration, the car exhibits Infiniti's DNA in a way that feels futuristic.

Infiniti • PROTOTYPE 9

"While the essence of the Infiniti Prototype 9 is rooted in the past, it runs on a next-generation EV powertrain, which looks squarely to the future."

Roland Krueger,
Chairman and Global
President of Infiniti

Infiniti • PROTOTYPE 9

The Power of "Wow"

The MOTOROiD may look like a two-wheeled robot, but it is more than that, being a careful consideration of harmony between rider and bike. Yamaha built this proof-of-concept experimental electric motorcycle aimed at creating new experiences of *kando*, the Japanese word for "deep satisfaction and intense excitement." We too have a word for this kind of experience: "Wow!" The prototype is equipped with extremely rapid balance control that senses its position in space, and can adjust its center of gravity via an Active Mass Center Control System (AMCES), which rotates parts such as the battery, swingarm, and rear wheel, using them as counterweights. This is a novel and awesome mashup of mechanical and electronic technology, and behind its design is surprising thinking: when you want to stabilize a vehicle on two wheels, you need to balance it, but if you don't, your bike should. Ultimately, this is not only a futuristic safety feature, it also enables a much closer interaction between rider and machine. No wonder the MOTOROiD is also able to recognize its owner. Pure *kando!*

Electro Swing

For many people a restoration project is a favorite pastime keeping their hands busy over the years. For Luca Agnelli it is both a labor of love and a real business endeavor. Trained in antique furniture restoration, he brings a very particular style to e-mobility, taking original vintage motorcycle parts such as fuel tanks, frames, forks, and coil-spring saddles and remixing them with batteries and rear-hub motors. The 36 V batteries are hidden in the "gas" tank; while an obvious hiding place, it still creates a completely different look than battery packs attached to the frame. Overall it feels a bit like putting a nineties beat over a swing classic from the forties, with everything played on twenty-first century instruments. This is not customization–it's really individualization. No two bikes are the same, but they all share the same personality. Agnelli chooses, observes, divides, and discards. He combines, coaxes, dismantles, and modifies, adapting to a bike's needs and studying shades of color, all with a sculptural approach. He is a curator of beautiful mobility.

Range
19–31 mi (30–50 km)
in relation to the pedal input

Battery
36 V 10.4 Ah

Torque
Bike models:
35 N m (25.82 ft-lb)
Kayak: 50 N m (36.88 ft-lb)

Weight
Bike models: 55 lb (25 kg)
Kayak: 88 lb (40 kg)

Top Speed
16 mph (25 km/h)

Availability
Built on request

Good design is timeless, and Luca Agnelli's e-bikes transcend their era, being universally desirable.

Not everyone is a vintage bike nut, but everyone loves Luca Agnelli's vintage-vibe e-bikes.

Mopeds from the 1950s are really cute, and beg the question: could Agnelli mass-produce?

His mix-and-match vintage parts, perfect chrome accents, and appealing color schemes are masterful.

Hiding the batteries in old gas tanks doesn't come off as wannabe, it just looks really cool.

A functional wooden kayak as an e-bike sidecar? Absolutely, and we'll take two, please.

Agnelli has an intuitive understanding for the exact right mix of leather, wood, paint, and chrome.

Agnelli Milano Bici

The Non-Buzzing Bee

A bio-based e-scooter—no steel, no aluminum, no carbon, just plants. There's also no frame: it's all a single structural shape. The monocoque body, made from flax (technically Dutch hemp—but don't call it weed) and bio-resin, provides strength and stability just like a monocoque in Formula 1. But not only that—it also replaces the traditional frame and its hundreds of separate parts (typically produced in an incredibly energy-sapping and time consuming process). You may have heard the adage "form follows function," but here "form follows material and production," as the designers put it. Beyond that, the silhouette of its shell and windshield makes for an unusual but very recognizable design. Developed by design studio Waarmakers in 2012, and later made possible by a crowdfunding campaign in 2015, it is planned to hit the roads in 2018, produced by Amsterdam start-up Van.Eko. The scooter features a 5.4 hp (4 kW) motor, a 2.5 kWh lithium-ion battery, and nice details such as the hand-stitched saddles (in vinyl or leather) and the huge integrated windshield, treated with a nanocoating for better visibility in rainy weather. Save the Be.es!

Range
50 mi (80 km) (at 45 km/h)
62 mi (100 km) (at 25 km/h)

Power
5.4 hp (4 kW)

Torque
110 N m (81.13 ft-lb)

Top Speed
16 mph (25 km/h)
28 mph (45 km/h)

Battery
2.5 kWh Li-ion

Weight
243 lb (110 kg) (incl. battery)

Availability
In production

Cape Not Included

Is it a bird? Is it a plane? No ...
It's a Bandit! The Vietnam-based motorcycle company that draws inspiration from comics and science fiction films gets a lot right and they explain this e-moto as being "from another dimension." This is not the first brand to take up bespoke production, but Bandit9 has turned madness into method, if you judge by details like burl-wood panels or Tuscan marble gas caps. The Odyssey is their signature bike, a true battle horse with an optional dual-drive electric engine that brings their forward-looking aesthetic where it belongs: into the future. The striking steel unibody conceals nearly everything technical, which includes performance components made from aeronautic materials. And the list goes on: Borrani rims, Marzocchi forks, Beringer Aerotec brakes, and a see-through LED display that is projected from below the tank and vanishes when off. This is a moving sculpture with a limited run of nine to be built. For collectors, or superheros?

Range
56–93 mi (90–150 km)

Power
54 hp (39 kW)

Torque
90 N m (66.38 ft-lb)

Top Speed
75 mph (120 km/h)

Weight
441 lb (200 kg)

Availability
Pre-production

Taking It to the Streets

Twin brothers Torsten and Bjorn Robbens are picking up where the Belgian factory Saroléa left off, with nearly 170 years of history as one of the hottest competitors in the European racing game, and a plan to do it all over again.

The expertise required for gunsmithing was a natural fit for building motorcycles, since casting, mass production, and precision machining are all required for good armaments. Saroléa's gun-making cousins included BSA, Royal Enfield, and FN, who all flourished in the first half of the century. Saroléa's first motorcycle, built in 1901, was a single-cylinder machine, basically a bicycle with a 381 cc four-stroke motor attached. But that humble start belied their later accomplishments. By the 1930s Saroléa was a very large factory, and developed the most formidable road racer on the Continent—the Mono-tube—which often beat more famous factories to the podium in the European Grand Prix. Special factory-backed Mono-tubes, ridden by professionals like Georges Monneret, gave the all-conquering Norton and Velocette teams consider-able grief, and period accounts remark on their difficulties against Saroléa when on the Continent.

100% electric, and blisteringly fast, as proved at the Isle of Man TT Zero, and soon, a street near you.

I

t is microscopic today, but the Belgian motorcycle industry has far deeper roots than all American and most European industries. Belgium's "Big Three"—FN, Gillet, and Saroléa —were located in the Herstal region, and in the first half of the twentieth century, they were major players in the European market. Saroléa was founded to make weapons in 1850 by Mathias-Joseph Saroléa, in the town of Liège. By 1896 the company had started building trikes thanks to the founder's visionary son, Mathieu-Joseph, who devoted the company's metallurgical skills to building motorcycles by 1901. Now twin brothers, Torsten and Bjorn Robbens, have resurrected the Saroléa brand that once built the most stylish, fiercely com-petitive racers this side of Italy. Saroléa aims to own that posi-tion once again, building white-hot competition e-motos that are available for street use too.

Belgium's "Big Three" – Saroléa, FN, and Gillet – all hail from the Herstal region, near Liége.

The Robbens brothers are picking up that thread, building hotrod e-bikes with monster performance and terrific style. They are the sole owners of the Saroléa brand, are 100% self-funded, and have been riding bikes together since they were four years old. Torsten spent 10 years working in F1 and Le Mans endurance racing. In fact, he's still the youngest team manager to win Le Mans, with the Audi Japan Team Goh. It was natural that Saroléa was reborn racing, with Torsten designing their first machine, the Saroléa SP7, for the Isle of Man TT Zero. This was clean-sheet design, starting with no assumptions, except the idea that the form had to be functional. That said, Torsten admits to adding design references to bikes he loves from the 1950s, 60s, and 70s. The SP7 is part of the first wave of electric superbike racers that begged the question: can we have this for the road?

It didn't take long before Saroléa turned its attention to a road bike, and the process of series-producing a road-legal electric superbike began in 2015. Their team has doubled to 20 "super motivated guys and girls" working on the first series called the MANX7. Bjorn explains, "We hired Serge →

"Saroléa is privately owned by Torsten and myself: we are 100% self-funded."

The SP7 was designed with clean-sheet principles, and more than a hint of vintage café racer styling.

Rusak, a Belgian designer, who retained the essence of the racing SP7 and developed our new Saroléa design DNA: the clear and simple design and vertical lines refer to the presence of our battery pack, symbolizing a rupture with present motorcycle design. When designing the MANX7 we did not make any compromises. We knew that the final result, you would love it or you would hate it. But it would certainly not leave you indifferent."

With the huge battery pack required for sustained high speeds, everything on the MANX7 needed to be as light as possible to offset battery weight. The compromise on building with the lightest components is price: a feather-weight build means the MANX7 is expensive—double the price of an IC bike of similar performance. That is not a recipe for mass adoption, but it is acceptable for a boutique brand carving its own niche in the marketplace: high-performance e-motos capable of over 200 mi (322 km) on a single charge.

The MANX7 roadster was designed by Serge Rusak, with all of the racer's DNA intact: a true hyperbike.

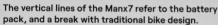

The vertical lines of the Manx7 refer to the battery pack, and a break with traditional bike design.

t's a whole different game," says Bjorn. "Our goal is to lead in terms of performance, efficiency, and safety, not in terms of volume. Taking the know-how and technology from our Isle of Man TT Zero motorcycles makes it possible for us to do so. At Saroléa all bikes are hand-built for each individual customer. Our customers are looking for something unique, something that has a soul and a story behind it."

The 170-year legacy of the Saroléa name is crucial to the brand's identity. "From the beginning we decided our motorcycles should last for generations. We still ride Saroléa bikes from the 1920s and 30s on a weekly basis. If we can make sure our MANX7 bikes are still riding in 50 or 100 years from now, we'll have had a positive impact: not needing to recycle something is the first way to go." That link with the past doesn't mean they aren't forward-looking; 90% of the high-tech components from the racing SP7 are incorporated in the MANX7. "With the hand-built monocoque carbon-fiber chassis, swingarm, and fairing, and the electric drivetrain with 120 kW of power and 450 N m (332 ft-lb) of torque on the shaft, I would say it's pretty exciting to ride." ●

An ultralight chassis was designed to balance out battery weight: the chassis is almost all carbon fiber.

"From the beginning we decided our motorcycles should last for generations."

Howling Wolf

In Norse mythology Fenris is the fearsome wolf that fights Odin, the king of the Norse gods, during an epic battle as the world ends. But, as with any good saga, the end is not the end but the beginning of a new chapter. Fenris Motorcycles, located in Denmark, has dedicated itself to developing electric superbikes for road and track that are fully homologated, street legal, and ready to race. And Fenris has been racing since 2014, when they competed in the MotoE series with early prototypes. The current proposed production bike incorporates a completely new chassis design adapted for electric motorcycles. The flat platform design, with hub-center steering, improves the position of the battery pack and increases driving dynamics. The platform system, common in the car industry, also enables easy construction of variants. The test bike shown here has been used for ergonomic riding and performance testing, but doesn't sport the finished design–so the myth remains untouched. That said, it is still vying to be leader of the pack.

Acceleration
0–62 mph (0–100 km/h):
2.9 sec

Top Speed
155 mph (240 km/h)

Range
155 mi (250 km)

Battery
14.4 kWh Li-ion

Power
200 hp (150 kW)

Availability
Launch 2020

Torque
250 N m (184.4 ft-lb)

Fenris Motorcycles • RL150RR

The Mean Green Performance Machine

Strong and stable: maybe not the most visionary idea in politics, but great to have in the rough and tough landscape of the Alps (or the dirt track closest to you). The Neematic FR/1 is an extreme-performance all-terrain electric bike from Lithuania. And as innovative as they are in the Baltics with digital ideas, they can be equally creative with analogue construction: the mid-drive motor of the FR/1 sits at the pivot point, minimizing wheel sprung mass and providing for a very low center of gravity. The two-part tubular-steel frame deflects at the correct point for top suspension performance and facilitating massive jumps, and the Fox rear shock comes with adjustable compression and rebound. You can choose between pedal-assist mode, with nine gears, or throttle control mode with up to 20 hp (15 kW) on tap. But the best part is the riding experience: the feel of an enduro motorbike, without the noise and fumes. Or, the other way around: the fun of going downhill on a mountain bike—without the hassle of getting back uphill.

Range
31–62 mi (50–100 km)

Power
20 hp (15 kW)

Torque
250 N m (184.4 ft-lb)

Top Speed
50 mph (80 km/h)

Battery
2.2 kWh Li-ion

Weight
115 lb (52 kg)

Availability
In production

Neematic · FR/1

Acceleration
0–62 mph (0–100 km/h):
9.6 sec

Range
186 mi (300 km)

Power
68 hp (4 × 12.5 kW)

Top Speed
91 mph (146 km/h)

Battery
110 V 200 Ah Ternary/
(NiCoMn) O2 21.9 kWh

Weight
1,797 lb (815 kg)

Availability
Production ready

The Czech Feather-Weight

The Czech Republic is a historic hub of automotive innovation, with a legacy of important technical advances since the 1800s. Today, over 150,000 Czech people work in car production, including for companies like Škoda, one of the oldest car companies in the world, founded in 1895. Thus, it is no surprise to see impressive innovations emerge from this wellspring of automotive knowledge. Take, for example the first modern production car to use four in-wheel hub electric motors; this was the well-known configuration of the Lohner-Porsche developed by Ferdinand Porsche in 1899. More than

100 years later, entrepreneur and designer Maurice Ward presents the Luka EV, benefitting from over a century of technological progress in lightweight construction, powerful e-motors, and smart battery solutions. The fiberglass body, inspired by the Tatra JK 2500 prototype from 1954, sits on an all-aluminum chassis, bringing the weight down to an impressive 1,797 lb (815 kg). That's less than a Karmann Ghia coupe from the 1950s— with more than double the power. The Luka EV is their very first car design; that bodes well for another century of Czech engineering innovation.

Like a Bolt from the Gods

Power
170 hp (127 kW)

Torque
393 N m (290 ft-lb)

Battery
14.4 kWh, Li-ion

Availability
Prototype

Nikola Tesla is not the inventor of the Tesla, Michael Faraday is not the CEO of a car company, and the late Glenn Curtiss did not design this bike. Instead, they were all influential engineers from the early Machine Age. Curtiss was a motorcycle and aviation pioneer, the godfather of the American V-twin engine, and is now a member of almost every technical hall of fame. In 2017 Matt Chambers founded the Curtiss Motorcycle Company, abandoning his former brand Confederate and announcing plans to build only electric motorcycles. The

company's vision is "a new golden age of American motorcycling—one based on sustainability, minimalism, and fun." Their first concept is the Zeus, boasting "the world's first E-Twin power unit," a set of two high-output electric motors driving a common output shaft. Designer Jordan Cornille grappled with post-IC motorcycle design in creating Zeus: if there's no fuel tank or motor to celebrate, how does the rider/consumer relate to this next generation of two wheels? Zeus is undeniably futuristic while retaining the design DNA Matt Chambers has produced since 1991.

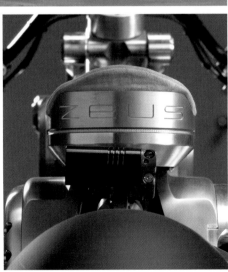

Taking a Gamble

Alta Motors's Team Built the Crapshoot, Their Second Semi-Factory Custom.

S

an Francisco's Alta Motors generates big buzz because they have big balls; they are the only e-bike factory pushing hard against the racing rules in place at both the FIM and AMA that bar electric bikes from competing head-to-head with IC machines. Alta has chosen the dirt path, building their extremely capable Redshift off-road racing motorcycles that have proven themselves in every condition. Since they are barred from officially sanctioned racing, they can only compete in privately sponsored events, where they always hold their own.

The success of Alta's Redshift platform is partly due to their founders' backgrounds: Derek Dorresteyn (chief technical officer) is a trained industrial designer, a former speedway racer, and counts two AMA Hall of Famers among his immediate family. Jeff Sand (chief design officer) spent years as an industrial design pioneer and invented the modern step-in snowboard binding at Switch. Dorresteyn is a vintage motorcycle enthusiast and earned his living with his own design and fabrication business, Moss Motors. While following the developments at Tesla Motors in the 2000s, he wondered: "If the tech is ready for a car, maybe it's ready for a motorcycle?" His riding buddy Sand agreed it was a thought worth exploring, and the pair sketched out some ideas. But they soon found that no available components matched their performance targets and they would have to build everything from scratch. By 2009 they had designed a motor that met their requirements and hired CEO Marc Fenigstein to helm Alta Motors. After a ton of testing, they began production of their Redshift motocross bike. The performance of this independently designed and manufactured motorcycle is exceptional, and when it's allowed to compete against gas-powered racers, →

> " The soul of motorcycling exists within the competition motorcycle: racers are pure. The marketer has been taken out."

Derek Dorresteyn

1960s drag racers ooze with style, and John McInnis took inspiration from twin-engined Triumphs.

Alta Motors • CRAPSHOOT

it often reaches the podium. But don't expect a Cinderella story full of guaranteed wins—racing is tough!

Designer John McInnis came to Alta to work under Jeff Sand after a stint at Lightning Motorcycles, where he shaped their 218 mph (351 km/h) World Speed Record e-bike, the Lightning LS-218. That same bike won the Pike's Peak Hill Climb in 2013 and sounded a gong in the motorcycle industry. McInnis now sketches concepts for Alta—he was tapped by Alta's marketing director Jon Bekefy to build a bike for the 2018 custom show circuit. The outcome was a silent dragster called the Crapshoot, and the press went wild for it. The Crapshoot has just enough traditional drag-bike styling to look vintage, but a quick motor-check reveals the Redshift heart underneath that fairing. "I planned on it having a hardtail from the beginning, and getting it as low as possible," said McInnis. He was inspired by 1960s dragsters like Boris Murray's twin-engine Triumph and Leo Payne's Turnip Eater: "These guys built fantastic machines out of garages in the 1960s, and I wanted to reflect that hand-madeness." The Crapshoot hits all the right notes and is among the first electric customs to excite even dinosaur-burning traditionalists by showing respect to the best builders of the past. ●

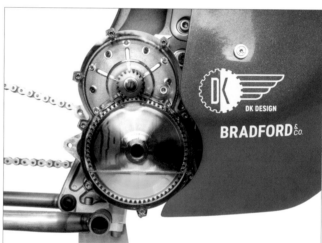

Range
50 mi (80 km)

Power
42 hp (31 kW)

Torque
52 N m (38 ft-lb)

Top Speed
80 mph (129 km/h)

Battery
5.8 kWh,
Waterproof Li-ion 350 V

Weight
275 lb (125 kg)

Availability
One-off

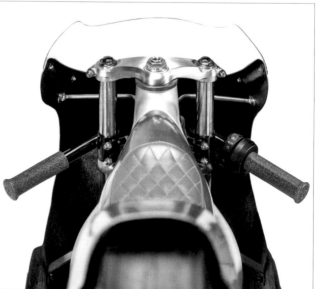

A fat drag slick, a padded seat/chest pad, and metalflake grips and footrests echo 1960s practice.

At its heart, the Crapshoot is an Alta Redshift in a sprinter's crouch: from dirt to dragstrip.

Crapshoot is a design exercise that proves electric bikes have an unlimited range of possibility.

"These guys built fantastic machines out of garages in the 1960s, and I wanted to reflect that handmade-ness."

John McInnis

Range
48 mi (77 km)

Torque
57 N m (42.04 ft-lb)

Battery
5.7 kW Li-ion

Availability
One-off

Power
28 hp (20.1 kW)

Top Speed
77 mph (124 km/h)

Weight
267 lb (121 kg) dry

Brushed Aluminum Beauty

Colt Wrangler, a cool-kid turned bike-builder, has some cowboy advice: "Look where you wanna go—it's the same with bulls and bikes." Colt would know, coming from a small town in Texas where he was born to rodeo-riding parents. Their son went on to found Colt Wrangler in 2015, which grew from a side-hustle into a custom bike manufacturer. His latest bike may finally prove that green is the new black; the Colt Wrangler Zero XU tracker is essentially a Zero XU from Zero Motorcycles, with Colt rebooting the bodywork to make it "look like a motor-bike," which meant adding a tank. He hand-shaped the aluminum—a first for him, as was his TIG welding work—and made the frame so sleek it weighs a mere 17 lb (7.7 kg). The faux tank makes for a polished look but also hides the electronics and extra converters that enable charging at Tesla Superchargers. Although wary of e-motorcycles at first, the nimble ride convinced Colt to turn this Zero into a raw-brushed street tracker. Shame it is a one-off by commission.

Colt Wrangler • ZERO XU TRACKER

Made to Measure

Italian Volt electrified the e-moto scene with the Lacama, a high-end, customizable electric motorcycle. The story of the Lacama starts in 2013, when two of Italian Volt's founders, Nicola Colombo and Valerio Fumagalli, captured the world record for the longest trip ever made on an e-motor-bike, riding 7,692 mi (12,379 km) from Shanghai to Milan. After their epic trek, Matteo Andreani, founder of the legendary Reunion race, commissioned them to build an e-moto for his race. The team brought on board their friend, designer Adriano Stellino, formerly of Lamborghini and Bertone, who brought his luxury-design experience to create the Lacama. Andreani wanted speed and he got it, with a top speed of 112 mph (180 km/h) and the trappings of any true tracker: knobby tires for gripping the dirt, and lots of power. Add to that a huge battery with a 124 mi (200 km) range, and what you get is an electric beast. But you might have to wait for it; each bike requires 24,000 hours of work, so the Lacama will be released as a limited run of only 20 pieces a year.

Acceleration
0–37 mph (0–60 km/h): 4.6 sec

Range
124 mi (200 km)

Power
94 hp (70 kW)

Torque
208 N m (153.4 ft-lb)

Top Speed
112 mph (180 km/h)

Battery
Up to 15 kWh; able to recharge quickly

Weight
540 lb (245 kg)

Availability
Limited edition starting in 2019

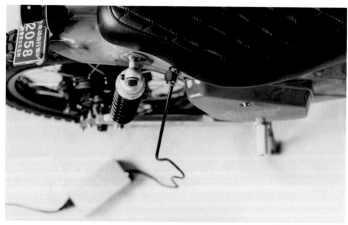

Range
25 mi (40 km)

Power
1.34 hp (1,000 W) rated
2.68 hp (2,000 W) peak

Torque
100 N m (73.76 ft-lb)

Top Speed
28 mph (45 km/h)

Battery
3.7 V 3.400 mAh Li-ion

Weight
176 lb (80 kg)

Availability
In production

The Retro Path Forward

It's nice to customize, but it's even more satisfying to build not only a bike, but also a brand. Shanghai Customs was established in 2014, importing Japanese bikes from the 1970s onwards, stripping them down and customizing them for sale. Since China has started to ban IC motorcycles due to pollution in large city centers, e-mobility is growing exponentially. Shanghai Customs reacted to this with a new interpretation of retro: they developed the eCUB 2, combining a Honda Super Cub chassis with an all-electric drivetrain that integrates a 1,000 watt electric motor in the rear

wheel hub and a bespoke removable battery pack. The result is not only satisfyingly "old school," it is also a step into the future of mobility. Since the Honda Cub is the most produced motorcycle in the world (with a production volume of over 100 million units in October 2017), the do-it-yourself kit by Shanghai Customs may see a very attractive future ahead. Shanghai Customs dreams of an "electro-sustainable lifestyle culture underpinned by a whole range of sustainable clothing and accessory options"—so your customized clothing can be as green as your bike!

Range
152 mi (245 km)

Power
9.7 hp (7.1 kW)

Torque
10.4 N m (7.38 ft-lb)

Top Speed
68 mph (110 km/h)

Engine
Air-cooled, 125 cc

Weight
298 lb (135 kg)

Availability
Built on request

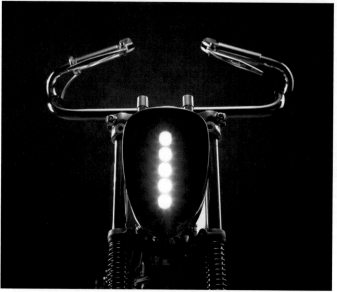

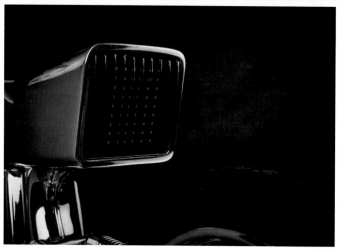

Never Exceed Warp 9

The bandits who brought us the Odyssey bike have no lack of ideas. Maybe because Daryl Villanueva, the founder and chief of design at Bandit9, started as an art director in advertising in LA; now he is building motorcycles in Vietnam. He has dubbed the L-Concept a "Sci-Fi masterpiece" and it is obvious why. The bike's body consists of a highly polished stainless-steel unibody tank and a suspended turbine-styled engine reminiscent of the USS Enterprise. The handlebars resemble chrome deer antlers, and the teardrop leather saddle is so subtle it looks like a mere dent in the tank. Notice the thin profile tires on the 21 in (56 cm) aluminum spoked wheels, and suddenly there are memories of something much older than a spaceship. The horizontal emphasis and the feeling of mounting a horse recall the very early days of motor-cycles—the 1885 Daimler Reitwagen, the world's first ICE-powered motorcycle, taken into the twenty-first century, or beyond. Whatever your associations are, this is a very unique bike.

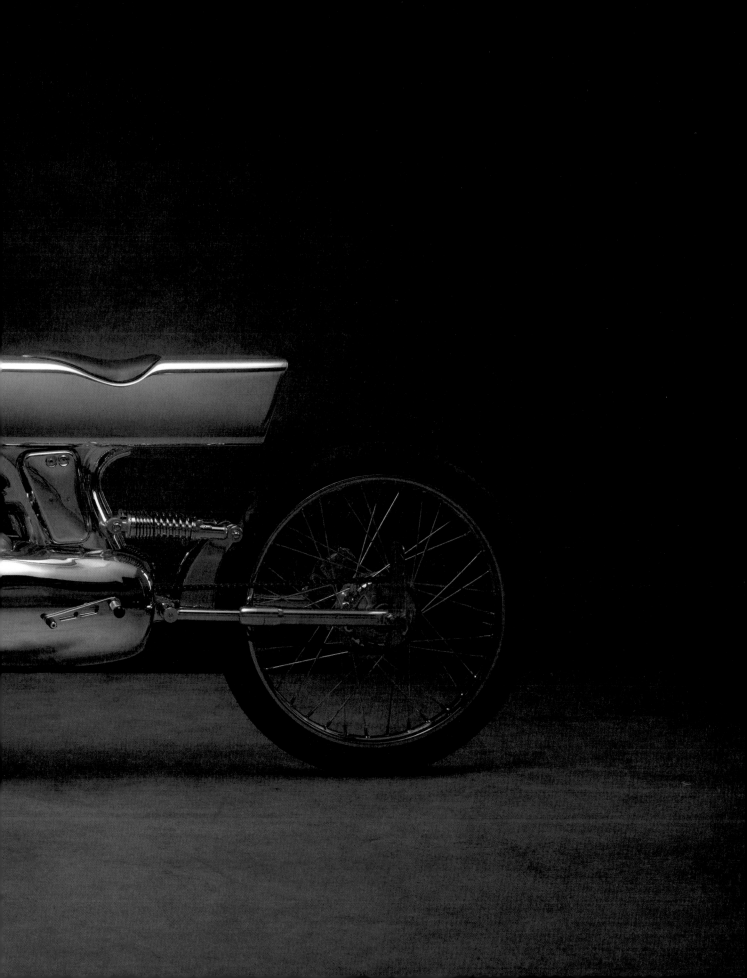

Go Anywhere, Do Anything

Lithium Cycles managed to capture every pubescent kid's dream: a cross-over of BMX style and motorbike feel with a retro minibike aesthetic. Powered by a 500 W (1,000 W optional) electric motor, the rigid-loop steel chassis looks at home on the sidewalk, the beach, or the mountains, and those ultrawide 4 in (10.1 cm) off-road tires give it a spirit of action and adventure. That motorcycle-style bench seat is comfortable, and because the Super 73 is a street-legal pedelec, you can ride it anywhere. Lithium Cycles was founded in 2011 in California to manufacture industrial carts and scooters, but in 2016 they put more fun into the mix: the Super 73 was born and has quickly grown into one of the most popular e-bikes on the internet, for all the right reasons. It comes in different flavors and colors, but fun is always the main ingredient. It just might be the cheapest way to feel like a kid again. Want added sound effects? This minibike was practically made for an ace of hearts to be clipped to the forks: brrrrrm!

Lithium Cycles • Super 73 SG1

Range
37–43 mi (60–70 km)

Power
250 watt nominal

Top Speed
16 mph (25 km/h)

Battery
696 watt hours / 48 V 14.5 Ah

Recharge
3–4 hours

Weight
70.5 lb (32 kg)

Availability
USA, Europe, Australia, Japan

Super 73 SG1

Super 73 Original 2016 model

Lithium Cycles • Super 73 SG1

Future Mother- Protector- Tutor

The BMW Vision NEXT 100 Reveals the Deep Core of Their Corporate Philosophy.

W

hat innovations in technology might transform the riding experience of the future? BMW's chief designer Edgar Heinrich pondered this question when given the challenge of bringing the motorcycle division's inspirations to BMW's Vision 100 program. While speculation about the motorcycles of the future has been popular for a century, the incorporation of those ideas into production is a legacy of commercial failure. Motorcycle designers walk a fine line of trying to push the boundaries of styling and technology while catering to a surprisingly conservative streak among the supposed rebels on two wheels. We want freedom and possibility, but we want it to look like what we already know. This paradox in our character is a conundrum for designers eager to adopt ideas like hub-center steering, feet-forward riding positions, or total rider enclosure—none of which are ever popular. This same effect stymies the development of electric motorcycles: as great an idea and riding experience as they are, experienced riders adopt electric bikes at a glacial pace unless they become the only option to ride, as we've seen in Chinese cities.

Luckily for Edgar Heinrich, an idea for 30 to 100 years from now is not subject to market trends or rider resistance, so new ideas can be explored freely. The most technically intriguing concept of the Next 100 Motorrad is the chassis, made from an intelligent mystery material that flexes, stiffens, and bends both to the rider's will and the road's demands. The frame is a triangular Möbius loop from axle to axle that supplants shock absorbers and even forks, bending itself to steer, changing from a light effort at low speeds to a stiffer push as the bike accelerates. The chassis, including the tires, intelligently responds to upcoming road conditions and the rider's speed, physically adapting tire →

Holographic displays, responsive goggles, and an intelligent suit that keeps the rider comfortable.

The chassis is flexible and incorporates suspension, and can stiffen or soften itself for the road.

BMW envisions the future-moto as partner and protector of its rider, with constant communication.

"I firmly believe the BMW Motorrad VISION NEXT 100 sets out a coherent future scenario for the BMW Motorrad brand."

Edgar Heinrich,
BMW Motorrad Head
of Design

tread shape and differentially stiffening parts of the frame itself, when necessary. The rider is kept informed of the motorcycle's suggestions on the appropriate line through corners via haptic signals in the handlebars and seat, as well as notes displayed on the (required) goggles, which double as information screens. The Vision Next 100's massive, soft-textured chassis is meant to project cocoon-like safety. And in fact, it employs intelligent systems so advanced the rider needs no protective gear, thus negating the helmets-leathers-boots-gloves mantra of our ATGATT (all the gear, all the time) culture of safe motorcycling. The machine will be more skilled than the rider, keeping itself upright and overriding rider error. Inevitably, the character and cultural assumptions of each manufacturer is embodied within its future vision: as a contrast, the Yamaha MOTOROiD is an erotic pleasure-enhancer, while BMW sees its role as mother-protector and tutor. The extrapolated future of motorcycling, as imagined by designers and manufacturers today, will see competing concepts defined less by styling or performance than by wildly different corporate philosophies concerning responsibility and rider interaction. ●

Daring to be Different

The Ujet Electric Scooter is a grand slam of high-design urban-mobility solutions all gathered into a single, unique vehicle with a host of clever features. It uses an asymmetrical ultralight frame made of magnesium alloys and carbon-fiber composites, and it is foldable. It also boasts the first-ever mass-produced spokeless orbital wheel with an in-wheel electric motor, and its tires are augmented with carbon nano-tubes. The future tech doesn't stop there: the onboard computer is app-controlled and integrates GPS, 3G, WiFi, and Bluetooth. It also has a touch interface for accessing the navigation, voice control, music playback, and an integrated HD camera. Did we mention there are integrated wireless speakers built into the battery—which also doubles as the seat? The battery comes in two sizes and can be easily charged from any power socket; it is removable, portable, and rollable, like your hand luggage. The total package is chic and futuristic, making a design statement that is among the most unusual two-wheeled vehicles seen in a generation. It follows none of the usual rules and cues for a scooter, yet still reads as one, and that is quite a feat.

Range
25–43 mi (40–70 km)
small battery;
50–87 mi (80–140 km)
large battery

Power
5.44 hp (4 kW)

Torque
90 N m (66.38 ft-lb)

Top Speed
28 mph (45 km/h)

Battery
48 V 2.3 kWh Li-ion

Weight
108 lb (49 kg)
small battery;
121 lb (55 kg)
large battery

Availability
In production

Forging a Path to Success

These vehicles not only look like they are from the beginning of the Machine Age, they are also built that way—in a proper smithy, run by the Bielawski family in Poland for over 100 years. Brothers Marcin and Michael have been interested in historical motorcycles and cars since childhood. Now they dream of producing retro electric vehicles of all styles, which means pedelecs are just the beginning. The brothers are using electric drives to propel their vintage designs and ancient methods into the twenty-first century. But the real showstopper here is the level of craftsmanship: a modular chromium molybdenum steel frame, steel handlebars custom-made for Kosynier, an aluminum tank concealing the batteries, and distinctive handmade leather accessories such as saddlebags and mudguards. And although it is family business, their marketing is cutting-edge; after ordering with them, you will receive a making-of video, filmed from six angles you can alternate between, showing the production process of your Kosynier Boardtrack or deLux bike. Be prepared to turn some heads.

Range
75 mi (120 km)
43 mi (70 km)

Power
0.33 hp (250 W)
1.34 hp (1,000 W)
5.4 hp (4,000 W)
6.7 hp (5,000 W)

Top Speed
16 mph (25 km/h);
31 mph (50 km/h) off-road

Battery
48 V 24 Ah Li-ion

Availability
In production

Artificial Balance

Artificial intelligence doesn't necessarily mean robots will take over the world, although they will soon mow the lawn, prepare dinner, and take care of grandma. AI will definitely be incorporated into motorcycles as well, in a bid to increase safety; this might might attract new users, or put off those who enjoy mastering a somewhat dangerous skill. One thing is clear: the question of how to balance such opposing needs remains open to interpretation. Honda's first foray into AI-supported motorcycling is the Riding Assist-e, an e-bike concept intended for novice riders. To start, this bike is impossible to drop at low speeds.

It also understands its speed and position in space and can take rider input into account, all of which help keep the learner safe. It has an extremely low seat height and low center of gravity, which makes beginners comfortable, and it has several riding modes available to match varying levels of skill and confidence. The question is: will they abandon the bike when they have learned to ride, or will the AI crutch remain a critical part of their skill set? That's the question of all AI-supported human activity, from driving to cooking dinner, and one we will be grappling with more and more in the future.

Making Monday Fun Again

What is a moped? The term derives from "motor" and "pedals", and historically, mopeds were just bicycles with a helper motor (like the famous VéloSoleX). In the 1960s and '70s they became hugely popular, especially with younger drivers; they were cheap, easy to use, and pretty fast (after a bit of tuning). Monday Motorbikes tries to cater to this feeling. Their M1 is a minimalist e-bike, originally based on a late-1970s Austrian Puch Magnum moped. It is built for city commutes, and its 7.4 hp (5.5 kW) motor delivers sufficient speed and power for most daily chores, or even a bit of fun. The 2.5 kWh 48 V quick-swap battery runs on a five-hour home recharge, or if you really want to, you could recharge it by pedaling while it is attached to the center stand. The ingenuity continues with a programmable keyless ignition, using a password or a specific sequence of button and lever movements, and it is Bluetooth-enabled for smartphone use. The M1 is designed and built in California and that's how it feels: it makes you want to drive and that's not a bad thing at all.

Range
40 mi (64 km)

Battery
2.5 kWh 48 V Li-ion

Power
1 hp (746 W) Eco mode /
7.4 hp (5.5 kW) Sport mode

Weight
170 lb (77 kg)

Top Speed
20 mph (32 km/h) /
40 mph (64 km/h)

Availability
In production

Monday Motorbikes • M1

GLOSSARY

Battery

The component that provides the power for electric motors in an e-vehicle. While electric cars, boats, and planes can carry sufficient batteries (and even solar cells) for long-distance travel, batteries remain the main stumbling block for electric two-wheelers, as none have been developed with enough power (see: kWh—kilowatt hours) for true long-distance riding. For decades, the rechargeable lead-acid battery was the foundation of EVs, but in the mid-1990s, new tech appeared: batteries using lithium-ion (Li-ion), nickel-cadmium (NiCad), lithium-polymer (LiPo). These new forms of battery were developed using NASA and industry research, and they now power everything from portable electric tools to Teslas.

Controller

An electronic system modulating the power delivery of an electric motor. Because electric motors deliver 100% torque at 0 rpm, a gradual feed of power is required for a smooth transition from standstill to road speed. The same is necessary for applying power during normal use. EVs draw tremendous electrical current, and controlling that current generates enough heat to cause fires; because of this, controllers on larger and/or faster vehicles use independent cooling systems. Controller software is a critical element in contemporary EVs, with the most sophisticated programs going into the most expensive vehicles.

e-bike

An electric bicycle that uses an electric motor as a supplement to pedal power.

Electric Motor

There are two main types of electric motor, AC and DC, with many variations on how they are constructed. All electric motors convert electrical energy into mechanical energy, but they vary widely in their power output. Some are built into wheel hubs (hub motors for e-bikes, e-scooters, and a few e-motos), but most are separate units bolted into a chassis. While an IC vehicle typically has one engine, EVs can have a motor near each wheel or within each wheel. There might even be multiple motors mounted within the chassis for more power. The more power a motor generates (from rider/driver demand), the more current it draws from the batteries, shortening its range.

e-moto

An electric motorcycle, wholly dependent on electricity sourced from the public power grid.

e-MTB

An electric mountain bike (i.e, bicycle), designed for off-road use.

e-scooter

An electric motorcycle with small wheels, usually with a step-through chassis design. E-scooters are the most popular EVs in the world, with around 200 million units produced since 2000, mostly for the Chinese market where small IC vehicles are outlawed in cities.

EV (Electric Vehicle)

Any vehicle relying solely on electric motive power: e-bikes, e-scooters, e-motos, plug-in cars, electric planes, boats, etc.

Hub Motor

An electric motor housed within the wheel of a bicycle, motorcycle, or car, typically as a central hub from which spokes and a wheel rim are attached. Especially common for e-bicycles, they are used in e-motos as well, and are sometimes installed in both wheels for two-wheel drive. The upside: no need for a chain, as the power is directly fed to the wheel. The downside: heavier wheels mean sluggish suspension response with more unsprung weight. It also means that "gearing" is fixed to the wheel diameter.

Hybrid

A vehicle that relies on an internal combustion motor to generate electricity, which is temporarily stored in batteries to run an electric drive system. The first mass-market hybrid-electric car was the 1999 Honda Insight, but the 2001 Toyota Prius was wildly successful. Hybrid motorcycles are still very rare.

IC (Internal-Combustion)

The principal type of motor used in vehicles today. An IC engine relies on burning fuel to create timed explosions inside a motor, which are then converted to motion via a series of crankshafts, gears, and shafts connected to the wheels. The first IC motors of 1814 used flammable powders, but the development of petroleum fuels in the 1880s led to the Otto (or four-stroke) engine, which is predominant in cars today. Other common IC motor types include diesel, two-stroke, Wankel, and jet engines. IC motors are much less efficient than electric motors, delivering only around 60% of their energy as motion (the rest is lost as heat), while electric motors are 97% efficient. IC motors use toxic fluids for fuel, lubrication, and cooling, and they also generate noxious gases, heat, and CO_2.

kWh (Kilowatt Hours)

The measurement of how long a battery's power will last—how long (how many hours) a battery will retain its power at a 1 kilowatt power draw. To illustrate: if a car only draws 1 kilowatt of power as it drives somewhere, its 14 kWh battery will last 14 hours. If, however, it draws 2 kilowatts of power to get somewhere, the same battery will only last 7 hours, and so on.

Pedelec

An electric-assisted bicycle that has no throttle: pedaling is assisted by an electric motor. It is generally believed Michael Kutter developed the first pedelecs in 1990, which he called the "Pedal Assist System." Yamaha built the first mass-produced pedelec in 1993, the YAS model, of which they have sold over four million units to date.

S-Pedelec

Short for Speed Pedelec which are not legally classed as bicycles since the power of the motor is < 250 Watts and the top speed exceeds 16 mph (25km/h) in Europe and 20 mph (32 km/h) in the United States. S-Pedelecs are usually classified as mopeds or motorcycles need to be registered and insured when ridden on public roads. A drivers license and wearing a helmet is required. In the United States S-Pedelecs are adopted into Class 3 category which limits the power to 750 watts and 28 mph (45 km/h).

INDEX

Morgan Motor Company
www.morgan-motor.co.uk
UK
Photography: Morgan Motor Company
Limited
pp. 26-29

MW Motors
www.mwmotors.cz
Czech Republic
Photography: Frantisek Nenutil
Additional credits: Maurice Ward
pp. 164-165

Neematic
www.neematic.com
Sweden
Photography: Aiste Kirsnyte
pp. 160-163

Night Shift Bikes
www.nightshiftbikes.com
USA
Photography: Harlin Miller
pp. 46-47

My Nobe
www.mynobe.com
Estonia
Photography: Seanest Ltd, Nobe Cars
pp. 10-11

Noordung
www.noordung.com
Slovenia
Photography: Luka Leskovsek
pp. 112-115

Onyx Motorbikes
www.onyxmotorbikes.com
USA
Photography: Steve Griggs (p: 88)
Steven Bianco (p. 89)
pp. 88-8

Porsche AG
www.porsche.de
Germany
Photography: Porsche AG
pp. 40-43

Rimac Automobili
www.rimac-automobili.com
Croatia
Photography: Rimac Automobili
pp. 74-75

Saroléa
www.sarolea-racing.com
Belgium
Photography: Jimmy Kets (pp. 152-155, 157)
Rob Mitchell (p. 156)
pp. 152-157

Shanghai Customs
www.sh-customs.com
China
Photography: Shanghai Customs
pp. 178-179

Shiny Hammer
www.shinyhammer.fr
France
Photography: Samuel Aguiar
pp. 16-17

Trefecta Mobility
www.trefectamobility.com
Netherlands
Photography: Trefecta Mobility
pp. 44-45

Ubco Bikes
www.ubcobikes.com
New Zealand
Photography: Wayne Tait (studio shots)
Jamie Gallant (p. 61, top)
pp. 60-63

Ujet
www.ujet.com
Electric Scooter
Photography: Ujet (pp. 195-196);
Patrice Meignan (pp. 194, 197)
pp. 194-197

United Nude
www.unitednude.com
Netherlands
Photography: Alex Lawrence
pp. 84-87

Vanderhall
www.vanderhallusa.com
USA
Photography: Drew Phillips
pp. 124-125

VanMoof
www.vanmoof.com
Germany
Photography: VanMoof
pp. 76-77

Vintage Electric
www.vintageelectricbikes.com
USA
Photography: Vintage Electric Bikes
pp. 90-93

Wannabe Choppers
www.wannabe-choppers.de
Germany
Photography: Juampi Garcia
pp. 64-65

Waarmakers
www.waarmakers.nl
Netherlands
Photography: Waarmakers
pp. 148-149

Yamaha Motor Company
www.yamaha-motor.com
Japan
Photography: Yamaha Motor Company
pp. 128-131, 140-143

ADDITIONAL CREDITS

Guillame Perraux's 1868 patent for an
electric bicycle motor
Copyright: Archives INPI
p. 2

Eugen Dutrieu's self-built 1898 electric
tandem pacing cycle
Copyright: Hockenheim Museum Archive
p. 2

In 1936 T. Hart installed batteries and a 12 V
motor into an ABC chassis,
a homemade cure for the Depression
Copyright: Francois-Marie Dumas/Moto-
Collection.org
p. 3

Mike Corbin's Quicksilver, shockingly fast at
165 mph, made possible by Yardney's NASA
batteries
Copyright: Mike Corbin
p. 4

Getty Images / Stephen Kim
p. 4, bottom right

Getty Images / Mirrorpix
p. 4, top right

The world's first e-superbike from the hand
of Yves Béhar and the team at Mission Motors:
150 mph (241 km/h) guaranteed in 2009
Copyright: Fuseproject/Yves Béhar/Seth
LaForge
p. 5

Electric Night Ride
Photography: Bob Van Mol
pp. 96-103

The Current

New Wheels for the Post-Petrol Age

This book was conceived, edited, and designed by Gestalten in collaboration with Paul d'Orleans

Edited by Robert Klanten and Maximilian Funk
Contributing Editor: Paul d'Orleans

Essay, Editorials, and Case Studies by Paul d'Orleans
Project Texts by Matthias Jahn (SOULD) & Nora Manthey (pp. 39, 60, 65, 67, 74, 88, 109, 126, 175, and 176)

Project Management by Sam Stevenson

Design, layout and cover by Hy-Ran Kilian

Typefaces: Roobert by Martin Vácha (Displaay Type Foundry) and Futura by Paul Renner

Cover photography by CAKE

Printed by Die Keure, Bruges
Made in Europe

Published by Gestalten, Berlin 2018
ISBN 978-3-89955-956-9

Paul d'Orleans is a journalist, curator, photographer, film producer, and expert on motorcycle history and culture. He publishes The Vintagent.com, and co-founded the Motorcycle Arts Foundation, supporting exhibitions, film, media, and events.

www.thevintagent.com
www.motorcycleartsfoundation.org

© Die Gestalten Verlag GmbH & Co. KG, Berlin 2018

All rights reserved. No part of this publication may be reproduced or transmitted in any form or by any means, electronic or mechanical, including photocopy or any storage and retrieval system, without permission in writing from the publisher.

Respect copyrights, encourage creativity!

For more information, and to order books, please visit www.gestalten.com.

Bibliographic information published by the Deutsche Nationalbibliothek.
The Deutsche Nationalbibliothek lists this publication in the Deutsche Nationalbibliografie; detailed bibliographic data are available online at http://dnb.d-nb.de.

None of the content in this book was published in exchange for payment by commercial parties or designers; Gestalten selected all included work based solely on its artistic merit.

This book was printed on paper certified according to the standards of the FSC®.